Airport Building Information Modelling

This book details how Building Information Modelling is being successfully deployed in the planning, design, construction and future operation of the Istanbul New Airport, a mega-scale construction project incorporating a varying mix of infrastructures including terminals, runways, passenger gates, car parks, railways and roads. The book demonstrates how Airport Building Information Modelling (ABIM) is being used to:

- facilitate collaboration, cooperation and integrated project delivery
- manage subcontractors and eliminate cost over-runs
- reduce waste on site and enhance overall quality
- connect people in a virtual environment to encourage collaborative working
- provide clients with an effective interface for lifecycle management including: design development, construction documentation, construction phases and BIM and Big Data Integration for future facilities management

The book presents a best practice BIM project, demonstrating concurrent engineering, lean processes, collaborative design and construction, and effective construction management. Moreover, the book provides a visionary exemplar for the further use of BIM technologies in civil engineering projects including highways, railways and others on the way towards the Smart City vision. It is essential reading for all Built Environment and Engineering stakeholders.

Ozan Koseoglu received his BSc and MSc in Civil Engineering & Construction Management from METU, Turkey, and his PhD in Construction Innovation and Technology Management from Loughborough University, UK. He worked for the largest private contractor in the UK on major projects in diverse sectors. He also used to work in the Middle East region on mega projects as a client representative with government and private sector ecosystems/authorities to manage large-scale investments across the lifecycle using integrated digital solutions. In his most recent role as CTO of Istanbul New Airport, he managed and delivered one of the world's largest-scope airport projects from design to operations in the Digital Environment. Recently, he established a global ecosystem called Digital Living Services for which he is Chairman and CEO with a track record of delivering 30 billion US dollar mega-projects globally. He is continuously involved in R&D studies and Technology Development Platforms for the future of Digital Living Services and has managed 30+ MSc and PhD students in many countries worldwide.

Yusuf Arayici is a professor at Hasan Kalyoncu University in Turkey, the dean of the faculty of engineering, and experienced research fellow of the Scientific and Technological Research Council of Turkey (TÜBİTAK). His research spans from Building Information Modelling and process modelling to energy efficiency and sustainability both in the UK and Turkey. He has led substantial research groups over a prolonged period of time through continuous cycles of research with funded research and enterprise projects summing £10.3 million, graduated thirteen PhDs and published more than ninety papers and four books in BIM.

Airport Building Information Modelling

Ozan Koseoglu and Yusuf Arayici

LONDON AND NEW YORK

First published 2020
by Routledge
4 Park Square, Milton Park, Abingdon, Oxon OX14 4RN
605 Third Avenue, New York, NY 10017

First issued in paperback 2023

Routledge is an imprint of the Taylor & Francis Group, an informa business

British Library Cataloguing-in-Publication Data
A catalogue record for this book is available from the British Library

Library of Congress Cataloging-in-Publication Data
A catalog record has been requested for this book

ISBN: 978-1-03-257051-8 (pbk)
ISBN: 978-1-138-32933-1 (hbk)
ISBN: 978-0-429-44824-9 (ebk)

DOI: 10.1201/9780429448249

Typeset in Bembo
by Swales & Willis, Exeter, Devon, UK

Publisher's Note
The publisher has gone to great lengths to ensure the quality of this reprint but points out that some imperfections in the original copies may be apparent.

I sincerely dedicate this book to Yusuf Akcayoglu, CEO of Istanbul New Airport, who had the great leadership, trust, vision, mentorship and management across the extraordinary delivery of the airport to become a worldwide benchmark for AEC and the aviation industry. He was fully on-board from day one to support BIM strategy alignment across the project with all parties. With his continuous support, friendship and inspiration for delivery, it turned out to be a significant success and benchmark for the future of the AEC industry.

Finally, this book is dedicated to my wife Selda and my little son Yigit Efe who have provided full encouragement and continuous support through my career and challenging delivery of Istanbul New Airport.

Dr Ozan Koseoglu

Contents

Preface

In the last two decades, several new technologies have emerged, by which our lives and works are surrounded. This, in turn, has become a stimulus for mankind to create new ideas and abandon the old ones. The more these technologies are used, the more our lifestyles and the way we work changes. For example, buildings and other civil structures were built in different ways depending on when they were built. Today, priority in design and construction is on the concurrency, synchronization, lean principles and integrity for the sake of sustainability.

Building Information Modelling (BIM) as a new and innovative methodology is at the forefront in the construction sector, where control of time, cost and waste is of paramount concern to all parties in the construction projects. Many problems relating to issues of control result from the inadequate communication of information within contracting organizations or amongst contracting and other design organizations. The amount of information involved in any construction project from start to finish should not be underestimated.

Although BIM is mainly considered for the building sector, this book expands the scope of BIM beyond buildings via its use in the Istanbul New Airport (INA) construction project, which is a mega-scale construction project incorporating a varying mix of infrastructures including terminals, runways, passenger gates, car parks, railways and roads. The book details how Building Information Modelling is being deployed in the planning, design and construction of Istanbul New Airport, and lays the foundation for its use in the Airport's future operation. Airport BIM is the outcome of extensive industry experience and academic know-how in the area applied as a best practice.

Authors came up with an idea of generating a flagship book for the future of digital design and construction in the AEC world. Therefore, Dr Ozan Koseoglu and Prof. Yusuf Arayici agreed to translate this valuable know-how and best practice into literature. This unique combination brought about the Airport BIM concept after 2 years of extensive work, following the grand opening of Istanbul New Airport into full operation.

The book presents a best practice BIM project with a new business model for clients while demonstrating concurrent engineering, lean processes, collaborative design and construction, and effective construction management. Moreover, the book provides a visionary exemplar for the further use of digital technologies in civil engineering projects including highways, railways and others on the way towards the Smart City vision. Accordingly, it demonstrates how Airport Building Information Modelling (ABIM) is being used to:

- facilitate collaboration, cooperation and integrated project delivery
- manage subcontractors and eliminate cost over-runs

- reduce waste on site and enhance overall quality
- connect people in a virtual environment to encourage collaborative working
- provide clients with an effective interface for lifecycle management including design development, construction documentation and construction phases, and also lay the foundation for BIM and Big Data Integration for future airport facilities management.

Acknowledgements

The authors wish to acknowledge those who have contributed to the *Airport Building Information Modelling* book:

- Elif Tugce Nurtan who has worked with the Istanbul Airport BIM team for endless days and nights with her full passion for success. She has delivered across the project with various roles on top of her MSc studies and successfully published academic work. She is currently in the UK and working on the next generation of digital construction.
- The Istanbul Airport BIM group and the whole project team who contributed to the delivery of BIM, which turned out to be a fascinating story of Airport BIM to share with the readers.
- Mehmet Sakin, who has been working on his PhD study related to the Istanbul Airport BIM delivery.

The authors give special thanks to Basak Keskin who has done the main editorial work of *Airport Building Information Modelling* from start to end with an amazing effort in the last 2 years. She has also done her MSc degree while she has been working for Istanbul Airport delivery with the rest of the team. She is currently pursuing her PhD study in New York for the next phases of airports, infrastructure and digital twins.

1 Introduction

CONTENTS

This book is about Airport Building Information Modelling. This is a term to deliberately highlight the use of Building Information Modelling as a new methodology for airports. Since Building Information Modelling is tightly linked with building construction, there is a prejudiced perception of BIM being useful only for buildings including existing or heritage buildings but not for civil engineering projects such as highways, railways or other infrastructure projects. Due to its nature, airport construction can encapsulate all the elements of construction types, e.g. buildings, railways, auto parks, and roads. If BIM can be used in airport construction, why not for civil engineering projects too? The term "Airport BIM" refers to the use of BIM for more complicated and sophisticated infrastructure projects, paving the way for BIM use beyond buildings.

If Airport BIM can be explained clearly using real data from a real airport project that is using BIM, it will then be possible to widen the spectrum for BIM use and mitigate the prejudiced understanding about BIM use being limited to buildings only.

Overall, the main purpose of this book is to identify how BIM can be used in airport construction projects to support design process, communication and coordination, concurrent design and construction practice, sustainable design and construction, construction planning and cost management, clash detection and timely procurement and effective facilities management. In addition, this book provides data and examples of a real case for BIM and Big Data Management. The holistic focus is to articulate how to utilize BIM throughout all phases of an airport project in design, construction and facility management. The chapters that follow clarify the substantial topics that should be considered in utilization of Airport BIM, and respectively these chapters are: "Airport Design and Construction", "Airport Building Information Modelling", "Concurrent Design and Construction with BIM", "Mobile BIM for the Airport Construction", "Key Learnings about ABIM and Paving the Way for Airport Operations", and "Conclusion".

The Istanbul New Airport project is briefly described in the next section.

1.1 Istanbul New Airport project

The Istanbul New Airport (INA) design and construction project has a scope encompassing 4 phases. The first phase has involved the construction of 3 runways, a terminal including 5 piers with an approximate area of 1.3 million m^2, a car park with an approximate area of 700,000 m^2 and other site facilities. The project is one of the largest investments in modern Turkish history and aims to provide significant economic improvements for the region.

The critical assets are the main terminal building, runways and related emergency runway/taxiway systems. At the end of the completion of all phases, a visionary project will come to life providing 76 million m^2 of airport with 6 runways, supporting 3,500 take-offs and landings per day, 200 million passengers a year, and access to 350 worldwide destinations.

In such a mega-size project, it is essential to track progress in accomplishing predetermined strategic and operational goals in terms of time and value management. Commission of Phase 1 of the project started in 2018 and this schedule date was the most important KPI for successful project completion on time. Since the aviation sector develops rapidly nowadays, efficiency in developing outstanding solutions to turn mega-scale investments/airports into reality within the planned/baseline schedule is critical for all stakeholders, especially for clients and investors. These airport investments are not just buildings but also create value for people.

1.2 Building Information Modelling (BIM)

BIM can be defined as the use of ICT technologies to streamline building lifecycle processes to provide a safer and more productive environment for its occupants, to effect the least possible environmental impact from its existence, and to be more operationally efficient for its owners throughout the building lifecycle. Therefore, a deliberation on the natural environment, user environment and owner satisfaction throughout the lifecycle is given within this definition.

BIM, in its most simple terms, is the utilization of a database infrastructure to connect built facilities with specific viewpoints of stakeholders. It is a methodology to integrate digital descriptions of all the building objects and their relationships to others in a precise manner, so that stakeholders can query, simulate and estimate activities and their effects on the building process as a lifecycle entity. There are a few definitions available for BIM in the literature as illustrated in Figures 1.1 and 1.2.

BIM can provide the required value judgements that create more sustainable infrastructures, which can satisfy their owners and occupants. However, it is necessary to realize that, while the users and owners may change over the lifecycle of a building within different intervals, the most important aspect is to minimize the impact to the natural environment.

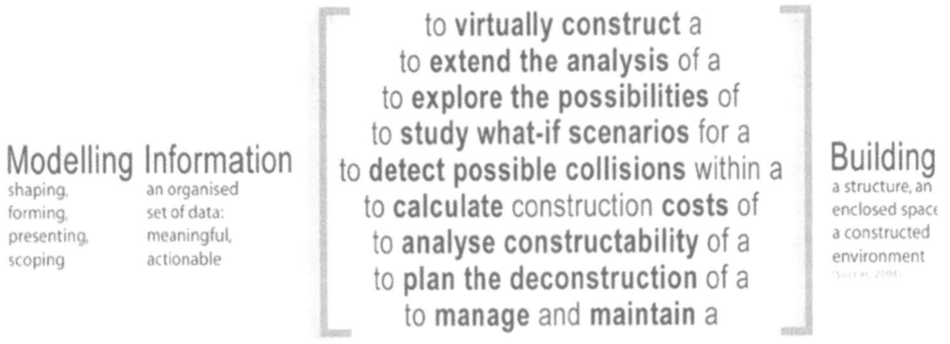

Figure 1.1 Words describing building, information and modelling (Succar, 2009)

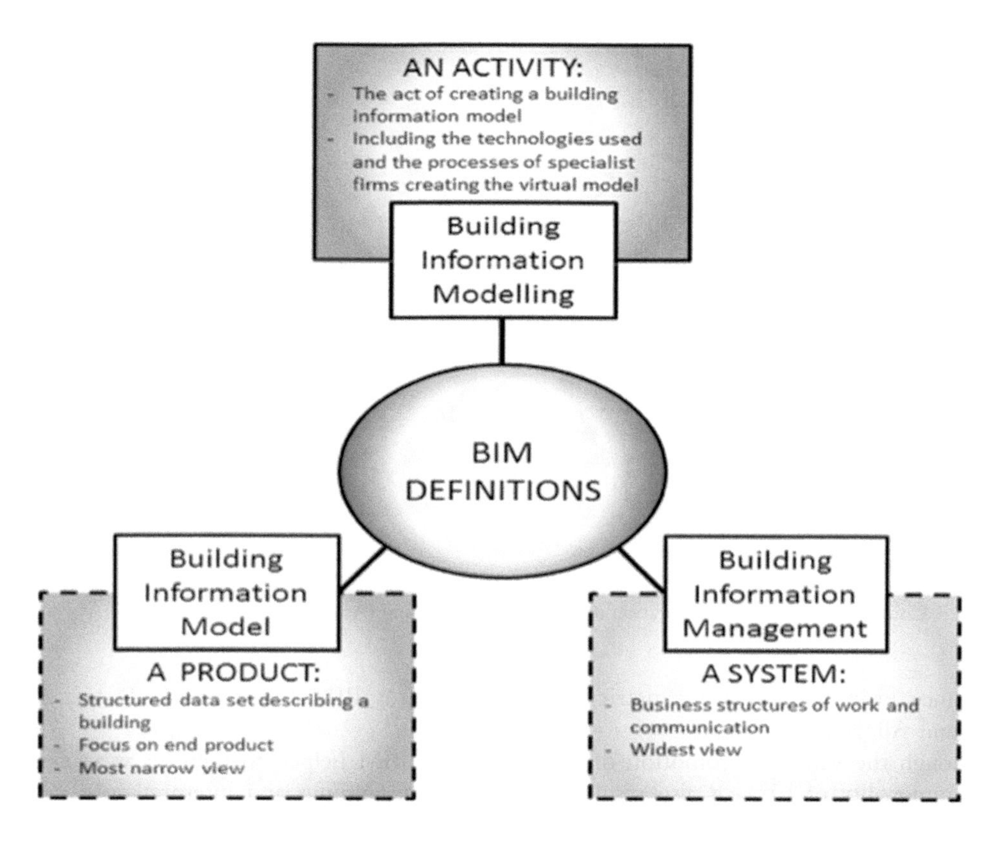

Figure 1.2 Different ways of looking at BIM (Maunula, 2008)

BIM as a lifecycle evaluation concept seeks to integrate processes throughout the entire lifecycle of a construction project. The focus is for stakeholders to create and reuse consistent digital information throughout the lifecycle (Figure 1.3). BIM incorporates a methodology based around the notion of collaboration between stakeholders using ICT to exchange valuable information throughout the lifecycle. Such collaboration is the answer to the fragmentation that exists within the airport construction industry, which has caused various inefficiencies. However, BIM could be the salvation of the airport construction industry since its use for concurrent design and construction in INA has provided strong evidence of BIM being a remedy for key challenges and issues that have remained unattended for far too long.

Concepts such as virtual BIM platforms have become critical since the 1990s and stimulated discussions, research and development for the elaboration of new powerful frameworks to support business models as illustrated in Figure 1.3.

CIC (Computer Integrated Construction) as a concept inspired by Computer Integrated Manufacturing (CIM) is a specific example of virtual enterprises for the construction industry as it aims to bind a fragmented and geographically spread set of construction stakeholders collaborating through the supply chain. It was sometimes called Building Product Models (BDP). However, contemporarily it is called Building Information Modelling. This is a term which originally emerged in the USA in the mid-2000s, and many CAD software vendors

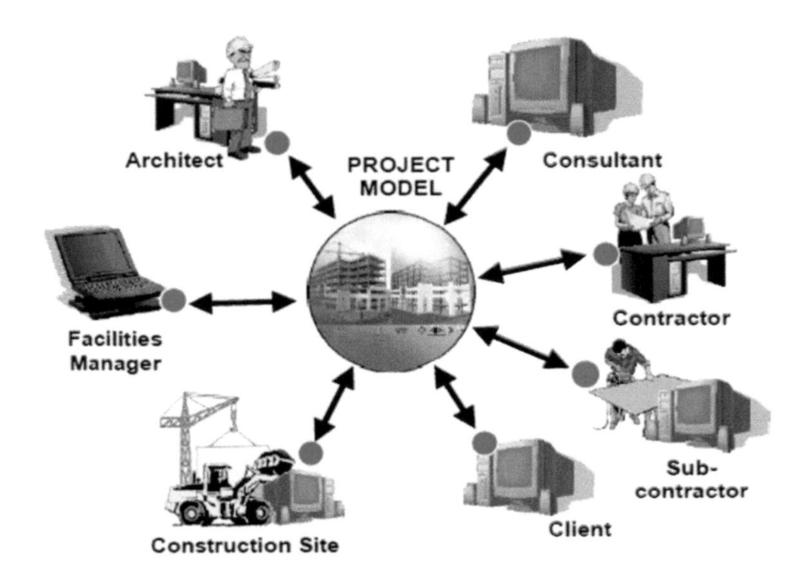

Figure 1.3 Communication and collaboration utilizing BIM modelling (Aouad and Arayici, 2010)

have since promoted their parametric modelling tools as BIM tools such as Revit, ArchiCAD and Allplan.

Although the vendors' promotion of the concept of BIM helped to increase awareness and the commonality of BIM, it also resulted in false understandings and interpretations of what BIM is amongst construction professionals. Furthermore, recent market and political pressures on the construction industry have led to a paradigm shift to (i) increase: productivity, efficiency, infrastructure value, quality and sustainability, and (ii) reduce: lifecycle costs, lead times and duplications, via the effective collaboration and communication of stakeholders in construction projects with a focus on the creation and reuse of consistent digital information by the stakeholders throughout the lifecycle.

1.2.1 BIM-based integration for a streamlined supply chain

Today, it is not uncommon to see a computer on every desk of an organization, linked together into a network. Through gateways, the communication can be global. Now different users can also run different applications on different hosts. BIM networks enable applications to exchange data information. For example, a designer might generate a design using a Macintosh PC in New York; the design analysis might be carried out in London, while the construction site might be in Tokyo, Japan. Where the information resides is not the responsibility of the users. The system itself should control the information management. The central core (with which applications can interact) can exist on one machine and other applications that run on a remote site can exchange information. For example, a CAD drawing may be generated in London, but the client in Istanbul can view the virtual reality model of the design without moving from the Istanbul office.

BIM acts as an integration framework encapsulating the backbone of the project data model and virtual communications can be possible via instant generation of 3D VR models. Its purpose is to integrate various construction applications into one integrated construction environment, where stakeholders work in harmony throughout the supply chain. The integrated environment enables orchestration through design, construction and the operational processes of the lifecycle.

Representation of the project lifecycle is not an easy task and may involve several experts; for example, architects, construction planners and site layout planners working on different areas. Therefore, it is useful to segment the data into a number of models, where each is concerned with a particular stage of the project lifecycle. This enables individual experts to work separately and makes it easier to manage data within the integrated environment.

1.2.2 Information sharing and collaboration

There is always a requirement for support tools to capture, in some way, the designer's intentions, going beyond a bland statement of shape and material. One of the aims of capturing that additional meaning is to prevent changes being made later in the design and construction lifecycle, and thus negating or lessening the value of design features.

Sharing enables the same feature to be used at multiple points in the structure. The designer can, therefore, augment the basic semantics provided by the product model. To take a simple example, where two objects are fitted together by means of a pipe join, the outer-diameter of one pipe must match the inner-diameter of the other pipe. If the common value is shared between both parts, a change made to either diameter will also apply to the other. Thus, the designer's intention that two parts should fit together is preserved.

The exchange of information is between the different computer systems and environments associated with the complete product lifecycle, including design, manufacture and maintenance. The information generated about a product during these processes is used for many purposes. It may involve many computer systems, including some located in different organizations. To support such uses, organizations must be able to represent their product information in a common computer-interpretable form that remains complete and consistent when exchanged among different computer systems. Furthermore, as being a fact in INA, there is a structural logic behind providing such information sharing and exchange processes that mainly focuses on enhancing the BIM end-user experience in terms of practicability and comprehensibility.

1.3 Rationale for BIM in INA

Investment in people was the key point for the INA project. Accordingly, sustaining the level of efficiency in project performance including Health and Safety was the top issue. A comprehensive strategic plan including worksite analysis standards and external training in hazard prevention and control measures are provided by the Department of Health & Safety in an extensive manner.

Likewise, engineering, quality assurance/quality check (QA/QC) and use of technology are also very critical. Therefore, from the very beginning of the INA project, a collaborative environment with subcontractors was vital and this also needed to be well coordinated on site too. This could only be achieved with Building Information Modelling on site. If all the engineers on site are fully equipped with technology, they can reach the current project models on the cloud platform to make a significant contribution to improved engineering quality.

There is also an environmental management plan that serves for the "green airport" scheme, which includes biodiversity action plan activities and the application of afforestation protocol; a wildlife management programme and usage of a bird monitoring and bird radar system. It is not only aimed to decrease the environmental effects during construction but also during the operational stage so an energy model should be developed for the whole infrastructure to ensure that the energy performance of the airport could comply with international standards such as the LEED accreditation. Thus, special tool and technologies should be integrated into the supply chain process to achieve the sustainability agenda.

In addition, control of time, cost and waste is of paramount concern to all parties involved in civil engineering projects. Many problems relating to issues of control result from the inadequate communication of information within contracting organizations or amongst contracting and other design organizations. The amount of information involved in any construction project from start to finish should not be underestimated.

At any stage of the project, different types of information are required by various people in various formats. For example, in large projects it has been revealed that more than 50% of site construction problems are attributed to design or communication of the design and more than 50% of contract modifications are related to design deficiencies. This suggests the need for early efforts by all participants to identify and resolve potential problems, ensuring delivery of complete and correct design and construction documents.

All in all, considering all these matters, using BIM in the INA project has been a vital necessity. BIM is elaborated in the next section.

1.4 BIM in the airport context: Airport BIM (ABIM)

The INA project has developed a strategy for utilizing designers, subcontractors and any other relevant parties and hosting all disciplines in a virtual collaborative work environment.

BIM strategy has been clearly defined by the client for the subcontractors and the client has been managing the whole process of integration on behalf of the subcontractors. Accordingly, the key subcontractors have strict obligations to follow and utilize the BIM process in their work processes because it has been predetermined that the BIM plan should lead the programme, cost plan quantities, and construction methodology while improving safety and reducing waste.

When the BIM execution plan was produced, a review and checking process analysed the capabilities of the designers, who were not familiar with BIM, and following this review process, they were then consolidated into the BIM-enabled integrated process. Figure 1.4 shows the BIM model of the Istanbul New Airport project, which acts as the central data model for the project.

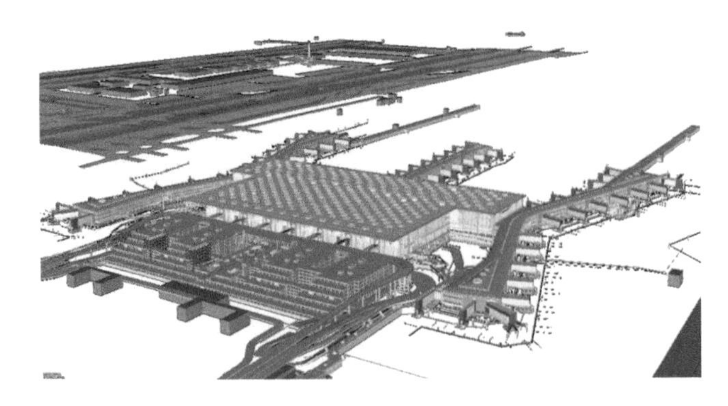

Figure 1.4 BIM model of Istanbul New Airport

A rendered view produced from the BIM model is illustrated in Figure 1.5, which helps for communication with the client and public. The rendered view is produced very quickly from the BIM model even though it is highly detailed.

In a holistic manner, according to the BIM plan, integration initially started with project designers using interface management procedures. A 30-month plan was developed that covered a lifecycle perspective including design, construction and operation stages respectively. A 4D model view is shown in Figure 1.6 to address the construction progress on site.

The BIM director of the project has led the development and execution of BIM implementation at all stages, starting from design and proceeding with construction and operation while managing the utilization of appropriate technological processes, cultural change and related human resources. Together with this, the BIM director has managed coordination between all elements of the airport's construction, including structure, architecture, MEP, baggage handling systems etc.

Figure 1.5 Rendering of the Istanbul New Airport project

Figure 1.6 An oblique plan view of the Istanbul New Airport BIM model

When considering the INA project's size, complexity and challenging duration, it is quite extraordinary. BIM has had a strategic role in executing engineering and design to accelerate the efficiency of design and construction, which has been a key driver for the team to be on time and even ahead of the construction on site.

BIM controls the subcontractors and eliminates any unforeseen cost, time and quality, which are of utmost importance for success. Herein, it is also crucial to know that success and benefits are gained via BIM, not because of how stakeholders come together through a virtual collaborative environment. That is why the key priorities are people and processes, followed by technology. Figure 1.7 shows the cloud system for BIM model sharing amongst the project team members to ensure on time progress and delivery on site.

Leading the project stakeholders including designers and subcontractors to familiarize them with the use of BIM in a coordinated fashion has been one of the important key points in this process of applying BIM in construction. Another important key aspect has been the designation of integrated project delivery (IPD) in a virtual room – also called a "BIM room" – that facilitates coordination for collaborative work and taking decisions with subcontractors and designers as shown in Figure 1.8.

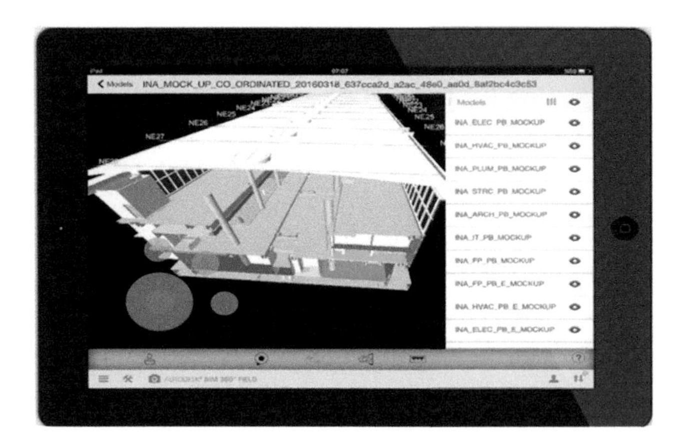

Figure 1.7 Product information cloud storage system of INA BIM

Figure 1.8 "BIM room" of INA

The outstanding part of this implementation is the way these key points were reproduced by 150 iPads including all coordinated BIM models used by the site engineers. Apart from 3D models, via the cloud system, approved 2D shop drawings were also provided for the field. Figure 1.9 shows BIM use on site.

Along with the production on site, QA/QC was also integrated into a cloud system and engineers of superstructure access information using their iPads, demonstrating concurrent design and engineering. Due to the INA project's size, number of stakeholders and amount of exchange and share, more than 30,000 activities for every day and month were integrated into the biggest 4D BIM model in the world, allowing real-time control over progress in the project. Figure 1.10 shows the use of tablets on site for progress checking via mobile BIM.

Figure 1.9 BIM on site with 3D renderings

Figure 1.10 BIM on site with tablets

Since the airport will be operated for 25 years after its completion, BIM will not only be used through the design and construction stages but will also continue to play a crucial role during the operation stage. Effective access to all necessary information and resolving future operational problems related to airport systems during pre-commissioning, commissioning and maintenance stages will all be possible with the usage of BIM. Additionally, the aim is to apply BIM at all dimensions including 6D of BIM. This covers the facility management/lifecycle management stage, which began after the completion of Phase 1 of the project, and has subsequently resulted in the provision of sustainability.

When the project duration was initially set and the scope and size of work for the project was identified, it was only possible to carry out and complete the project through concurrent engineering with BIM; it was clear that the INA project would lead and set the benchmarks for the AEC industry and digital construction. Consequently, the Istanbul New Airport project was selected as a finalist in the large infrastructure category of the Autodesk 2016 AEC Excellence Awards among 162 competitors from nearly 30 countries all around the world.

1.5 Conclusion

BIM is not a new concept anymore but it has just started to draw attention in the construction industry. Some governments are introducing legislations to make BIM use a mandatory requirement on public projects and some private clients are also paying attention to it. Many consultants and contractors have already started to utilize BIM tools and processes at company level by avoiding any direct contractual obligations to their client because they see the potential for cost and programme savings.

The adoption of BIM is gaining momentum in Turkey as the Istanbul New Airport project has become a key learning hub, in which enormous numbers of contractors, subcontractors and designers are managed and coordinated in the collaborative BIM environment. Besides, this has already provided some evidence and warranty to overcome the limits that the Turkish construction industry has been facing. Furthermore, replacing conventional construction industry methods that have been used for years by introducing a developing complex know-how has been the most challenging part in the adoption process. Cultural change can only be possible with a clear vision that defines strategy and an effective BIM implementation plan. That is why only having a grasp of the theoretical background of BIM is not enough, defining the clear vision, outlining the right strategy and seamless execution are crucial and essential too.

Bibliography

Aouad, G. and Arayici, Y. (2010) *Requirements Engineering for Computer Integrated Environments in Construction*. Oxford, UK: Wiley-Blackwell.

BIM Task Group (2012) Building Information Modelling; Industrial Strategy: Government and Industry Partnership, www.bis.gov.uk

Khosrowshahi, F. and Arayici, Y. (2012) Roadmap for implementation of BIM in the UK construction industry, *Engineering, Construction and Architectural Management*, Vol. 19, No. 6, pp. 610–635.

Maunula, A. (2008) The Implementation of Building Information Modeling (BIM), A process perspective, Helsinki University of Technology SimLab Publications, Report 23.

Succar, B. (2009) Building information modelling framework: a research and delivery foundation for industry stakeholders, *Automation in Construction*, Vol. 18, No. 3, pp. 357–375.

2 Airport design and construction

CONTENTS

2.1 Introduction

Airport projects are highly complex in terms of design, construction and operations of a varying mix of infrastructures including terminals, piers, runways, taxiways, hangars, car parks, railways, and roads. They have significant socio-economic impacts by transferring and connecting people, goods and freight. The Vision 2050 Report by the International Air Transport Association (IATA) (2011) notes that the growth of aviation will double the rate of the global economy in the next 25 years, with the number of passengers 1.4 times higher than today. Increase in the number of airlines and hub airports leads to a more competitive environment in the aviation sector. At this point, the quality of airport design and construction comes into focus.

After World War II, airport design became more refined as supply and demand for air transport infrastructure increased significantly. Starting from a decentralization trend that includes the use of piers, fixed linked bridge and jet bridge systems, generics of airport design have transformed substantially. Increasing and changing demands and approaches to airport design and construction have shaped today's modern airports (Uffelen, 2012). Modern airports, which are also called "airport cities", do not just offer terminal and runway operations, but also car parks, logistics, lounges, malls, hotels, retail areas, railway stations and conference halls too.

While the Istanbul New Airport project is planned to be one of the world's largest aviation hubs, it also provides a unique airport city experience by utilizing its geographically advantaged position at the crossroads of Asia, Europe and the Middle East with facilities including:

- Terminal building with an approximate area of 1.3 million m^2, and unique architecture that mainly represents the cultural heritage of Istanbul
- Cargo city encompassing the warehouses, agency buildings, customs office, service points (banking services, cafes and restaurants, dry cleaning, hair dresser, PTT (postal mail service), prayer rooms, veterinary centre, medical centre and test laboratories)
- Retail and Duty Free on an area of 55,000 m^2

- Airport hotel in the main terminal building, which aims to be the largest airport hotel in Europe
- Airport car park with a total capacity of 40,000 vehicles
- Subway and speed train
- Port facilities for fuel services

Conceptual design renders of the Istanbul New Airport project can be seen in Figures 2.1 and 2.2 from different visual perspectives and styles.

It is obvious that the design and construction phases of INA are very challenging due to the aforementioned capacities of the airport city facilities. On top of the targeted size of the airport, as a consequence of the project delivery method used for INA, which is build-operate-transfer (BOT), the predetermined project duration has also added an extra layer of difficulty to the project. Overall, all project stakeholders have been confronted by both technical and managerial challenges in the design and construction executions of the project.

For instance, from design, engineering and construction perspectives, coordination of MEP (mechanical, electrical and plumbing) systems and BHS (baggage handling systems) has been one of the major challenges. To give another example, the high number of different parties (subcontractors, designers, vendors, client etc.) has also led to immense challenges in the communication flow, especially considering the size and complexity of the project.

There are also other projects similar in nature to INA that underwent the same challenges, as those projects were also aiming to be hub airports in their own geographies. Since such airports generally reflect great cultural and socio-economic value, they are considered as signature iconic projects that are architecturally attractive and eye-catching. Single roof canopies, an abundance of steel structures, green roofs, articulated façades, glazed openings, skylight apertures, pools, passive systems and three-dimensional representations can be listed as some of the preferred architectural features of those iconic airport projects. Creating an innovative model of engineering, architecture and sustainable design is of utmost importance for airport projects.

Figure 2.1 Bird's eye view of INA[1]

Figure 2.2 Conceptual design render of INA[2]

Experience from past projects has contributed to the development of the INA design and construction management strategy, in which BIM has played a crucial role. Digital design, engineering and construction processes have been the key approaches to tackling the aforementioned challenges due to the complexity, ambiguities and uncertainties in the INA project.

Today's modern airports choose to offer the following smart technological solutions to provide an advanced passenger experience:

- IT solutions including A-CDM (Airport Collaborative Decision Making)
- Facility management with BIM (Building Information Modelling) and VR (virtual reality)
- Augmented reality
- Smart kiosks
- Social media-enabled services
- Airport gaming
- Loyalty programmes
- Queue management
- Airport mobile application

2.2 Key issues on airport design and construction

From the outset of the 21st century, people's needs and requirements have been hyper-evolving and there are many issues to be considered for complex system developments such as airports. For example, capacity, aircraft and airport compatibility, sustainability and technology aspects can be listed as the key concerns in airport design. A prerequisite of successful airport operation is a good design solution. The facilities listed below are high-level basics of airport design:

- Runways
- Runway strips
- Taxiways
- Aprons

- Pavements
- Aircraft ground handling
- Aircraft refuelling
- Cargo
- Passenger terminals
- Security areas
- Landside access
- Visual aids for navigation
- Utility centre
- Air traffic control tower

Basic airport systems can be spatially divided into two groups: landside and airside. There are many complex systems associated with these two spaces. Figure 2.3 depicts the ecosystem of the airport building and infrastructure systems, which are all coordinated in the BIM environment. The BIM master model encompasses all the major structures that belong to airside and landside regions. In addition, Figure 2.3 provides a high-level overview of how different parts of airport infrastructure are categorized in terms of the zones they are associated with. The assets defined in between airside and landside zones are also crucial as they hold significant complexity in terms of coordination between superstructure and underground systems. To elaborate on their interaction, the characteristics of each given structure are as follows:

Terminal: This is the core building where passengers transfer between ground transportation and the airport facilities to board and disembark from the aircraft. The terminal is an interface between the public space and the airside zone. In the terminal area, ticketing, checking in and security screening processes are deployed. In the INA case, at the end of completion of all phases, there will be multiple terminals with multiple concourses which will be connected through walkways, sky-bridges, or tunnels. Phase 1 encompassed only Terminal 1 which has a total area of approximately 900,000 m^2. Design and construction of Terminal 1 has now been completed.

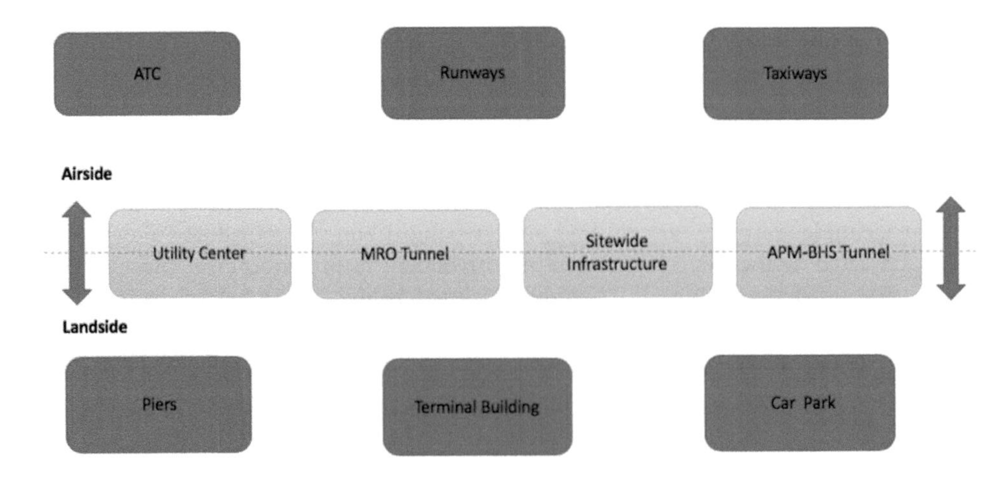

Figure 2.3 INA airside and landside structures

Piers: Piers are part of the terminal building. They offer high aircraft capacity. In the INA project, pier finger terminals were incorporated in the design. There are five piers in total offering a total area of approximately 320,000 m^2.

Car park: In the INA project, the multistorey car park is designed to have open spaces – atriums. The total car park area is approximately 700,000 m^2.

Airport traffic controller (ATC) tower: ATC tracks each flight from departure to arrival to maintain safety on airside. In the INA project, the ATC tower stood out from among 370 projects to win the International Architecture Award. This building holds iconic value for the airport.

Runways: In the INA project, there will be three runways in total for the Phase 1 as can be seen in Figure 2.4.

Taxiways: Taxiways allow aircrafts to travel between the runways and other airside zones.

Utility centre: This has the purpose of maintaining the infrastructure required for airport operations. It includes extensive networks of cables, pipes and installations.

Maintenance, repair and operations (MRO) tunnel: As the name implies, it is a hub for any activity related to repairs, adjustments, replacements, inspections, rebuilding etc.

Airport people movers (APM) and baggage handling systems (BHS) tunnels: These tunnels provide automated transit services for people and baggage.

A master model depicting the structures listed above is shown in Figure 2.4. This master model belongs to Phase 1 of the INA project.

Due to the scale and operational complexity of the INA project, design coordination has been a great challenge. The subsystems of each landside and airside structure that had to be coordinated are given below:

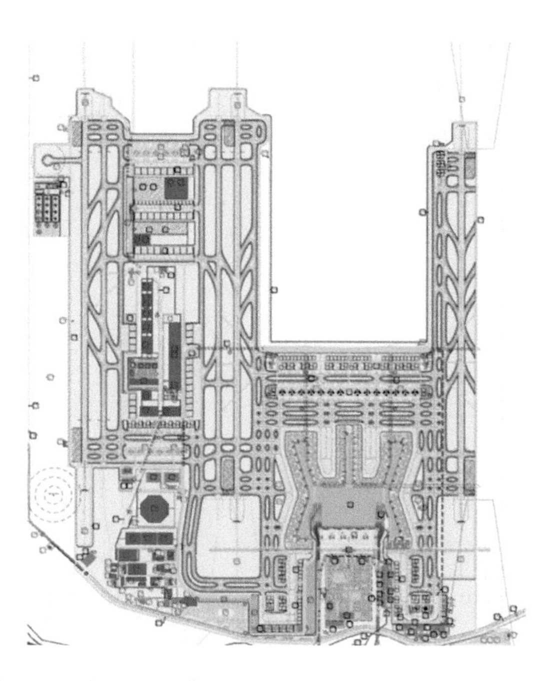

Figure 2.4 INA 1st initial phase sample master plan

Terminal building

- Architectural and structural systems
- MEP systems:
  - HVAC ducting
  - HVAC piping
  - Plumbing
  - Fire protection
  - Electrical system (cable trays, ducts)
- BHS systems:
  - Conveyor
  - BHS steel
  - Cable trays and ladder

Piers, ATC, car park, utility centre

- Architectural and structural systems
- MEP systems:
  - HVAC ducting
  - HVAC piping
  - Plumbing
  - Fire protection
  - Electrical system (cable trays, ducts)

Runways

- Surface models
- Airside drainage
  - Open channels
  - Culverts
  - Filter drains
  - Slot drains
  - Manholes
  - Pipes
- AGL main infrastructure
  - Galleries
  - Primary duct banks
  - Manholes
  - Handholes
- Navaids main infrastructure
  - Primary duct banks
  - Manholes
  - Handholes

- Bases and foundations
    - Substations
    - Floodlighting
    - Radar and NavINAtional aids
    - Blast barriers

Site-wide infrastructure

- Underground networks
- Fuel hydrant
- Fire hydrant
- Storm water
- Water supply
- Potable water
- Grey water
- Natural gas
- Irrigation
- Waste water

Aforementioned airport MEP-IT and infrastructure network systems that belong to ATC, airside and terminal areas are visualized respectively in Figures 2.5, 2.6 and 2.7. As depicted, the BIM master model of such a mega-scale airport design and its construction phases has tremendous complexity.

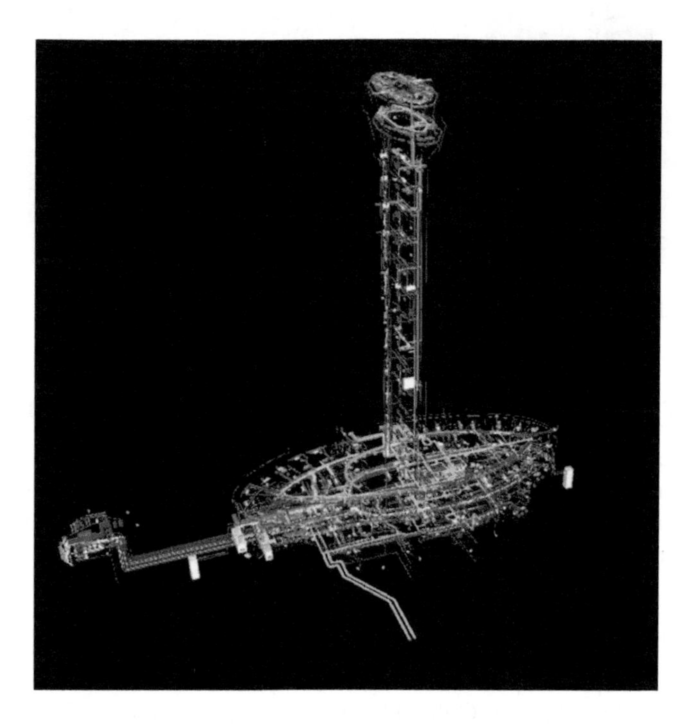

Figure 2.5 Coordinated MEP-IT systems of ATC

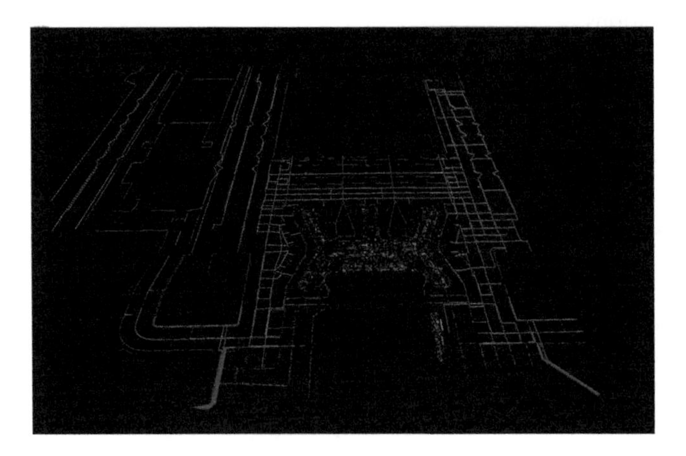

Figure 2.6 Coordinated infrastructure network systems

Figure 2.7 Coordinated terminal MEP-IT systems

2.3 Examples of airport design and construction around the world

There are good examples of airport design and construction around the world that offer key lessons and experiences which can be utilized for improved design and construction performances to address complex needs and requirements. Some iconic projects around the world and their design and performance overview are elaborated below.

2.3.1 Kuwait International Airport Terminal expansion

Kuwait International Airport is planned to significantly increase its available capacity and provide a new regional hub in the Gulf. The terminal has a trefoil plan, comprising three

Figure 2.8 Render of Kuwait International Airport[3]

symmetrical wings of departure gates (see Figure 2.8). To aid orientation, the building is planned to have a single roof canopy, punctuated by glazed openings that filter daylight, while deflecting direct solar radiation (Uffelen, 2012). Approximate numbers for total area and capacity increase are as follows:

Total area increase: 708,000 m^2 to 750,000 m^2
Capacity increase: 13 million passengers per year to 25 million passengers per year

2.3.2 Abu Dhabi International Airport Midfield Terminal Complex

Abu Dhabi Airport Midfield Terminal Building aims to be a benchmark in airport construction worldwide as one of the most architecturally impressive structures. Design of the airport includes an undulating roof, inclined façade, 52 m processing ceiling, and 3D and also 4D virtual collaborative environment (see Figures 2.9 and 2.10).

Figure 2.9 Render of Abu Dhabi International Airport[4]

Figure 2.10 Aerial view of Abu Dhabi International Airport construction[5]

A central source of approved data was used by project team members to undertake their work such as material take-offs, programming and the development of fabrication drawings. The project team aimed at savings from the BIM use. They also planned to reduce critical RFIs (Requests for Information) via effective BIM use. The terminal was due to open in September 2017, yet it was delayed for 2 years until 2019 due to the complexities that the project team has encountered in the design and execution processes. Approximate numbers for total area and capacity are as follows:

Total area: 742,000 m^2
Total capacity: 27 million passengers per year

2.3.3 Heathrow T5

The construction of the fifth terminal at London Heathrow was the subject of a limited competition in 1989. The design included large bands of glazing to make great use of daylight, and also big structural trees to support the roof and enhance the structural design (Uffelen, 2012) (see Figures 2.11 and 2.12).

Figure 2.11 View of Heathrow T5 construction site[6]

Figure 2.12 Render of Heathrow T5 interior design[7]

From a design perspective, a number of aspirations were set with the intention of supporting BAA's goals. Core to these aspirations was the single model environment (SME), which was subsequently adapted to the common data environment (CDE) to recognize an even broader purpose. The SME/CDE brought many benefits. One of the most important benefits was asset integration: T5 aimed to have a smooth data flow from concept through to completion and on into maintenance. Approximate numbers for gross floor area and capacity are as follows:

Gross floor area: 465,000 m^2
Total capacity: 30 million passengers per year

2.3.4 Beijing Capital International Airport, T3

Beijing Airport is one of the world's biggest airports and was completed for the 2008 Olympic Games. The aerodynamic roof evokes traditional Chinese colours and symbols and symbolizes the fluid motion and thrill of flight (Uffelen, 2012) (see Figures 2.13 and 2.14).

Approximate numbers for total area and capacity are as follows:

Figure 2.13 Render of Beijing Capital Airport interior design[8]

Figure 2.14 Aerial view of Beijing Capital International Airport[9]

Total Area: 1,300,000 m^2
Capacity: 50 million passengers by the year 2020

The terminal building is one of the world's most sustainable, incorporating a range of passive environmental design concepts, such as the south-east orientated skylights, which maximize heat gain from the early morning sun, and an integrated environment-control system that minimizes energy consumption. Overall, from a sustainable design perspective, Beijing Airport is a really good example of smart infrastructure.

2.3.5 New International Airport for Mexico City

Mexico City's New International Airport will be the biggest in North America and the third-largest in the world (see Figure 2.15). It is the largest infrastructure in Mexico's history, and was conceptually initiated in 2014. The total estimated cost is 14 billion US dollars. Design considerations target LEED Platinum certification as the entire building is serviced from beneath, freeing the roof of ducts, pipes and revealing the environmental skin. This hardworking structure harnesses the power of the sun, collects rainwater, provides shading, directs daylight and enables views, all while achieving a high performance envelope that meets high thermal and acoustic standards. Hence, BIM is planned to be a key player in the project for achieving sustainability objectives and improving design decisions.

The documentation for the terminal building was similar in scale to an urban-planning exercise. The team directly created from the model more than 10,000 construction documents and more than 5,000 room data sheets for each room in the buildings. It is estimated that they would have had to hire 1.5 times more staff to produce the same amount of drawings without BIM (Autodesk, 2017). Approximate numbers for total area and capacity are as follows:

Total area: 743,000 m^2
Capacity: Initial stage 75 million passengers per annum – when fully completed 125 million passengers annually by 2022

Figure 2.15 Render of New International Airport for Mexico City[10]

However, according to the several news sources, the project was suspended when the airport was one-third complete. There is argument on an alternative plan to convert a military airbase into a commercial airport instead of completing the cancelled project (Reuters, 2019).

2.4 Planning, design and construction at the Istanbul New Airport project

Being one of the world's largest airport projects, the project scope encompasses 4 phases. The first phase has involved the construction of 3 runways, a terminal including 5 piers with an approximate area of 1.3 million m^2, a car park with an approximate area of 700,000 m^2 and other site facilities. In project scope, the critical assets of the main terminal building, runways and related emergency runway and taxiway systems will be achieved.

INA has been planned to be constructed and operated for 25 years. The aim is to provide a global hub for aviation to enhance Turkey's commercial and tourist potential. The biggest challenge to implementing connected BIM on such a large-scale project in Turkey was the cultural change. Transforming the way people collaborated and getting all of the stakeholders to embrace new technology into their processes was key for this complex project. The success of the project has been a major driver of change and integration for the Turkish construction industry. The long-term goal for the future of this unique and complex facility is to integrate construction information and operational data using connected BIM as the main management platform across the lifecycle of this project, providing huge benefits in time and cost savings over the next 25 years. Utilizing a connected digital environment across the lifecycle of the project was vital in bringing this complex project to completion on time and within budget. Being able to work on design, engineering and construction processes in parallel provided significant time savings and added confidence in bringing the airport into operation quickly and with high quality.

In Figure 2.16, a construction site photo of the project's first phase, which was nearing completion at the time, can be seen. Integrating the dataflow of the design and engineering phases helped eliminate issues and discrepancies before construction began. Carrying the utilization of BIM environments forward into the operational phase of the facility, the Istanbul New Airport

Figure 2.16 Aerial view of INA Phase 1 construction site[11]

project will continue working with BIM data for real-time evaluation and integration of the lifecycle requirements of the airport's many systems.

Utilizing connected BIM from construction to operation with the right skillset and people transformation provided a very quick return on investment for making the airport operational. Grasping BIM as a transformative innovation process was significant in this journey. An integrated project delivery mindset was also part of the BIM implementation process. Accordingly, the journey started with design including the steps of conceptualization, criteria design and detailed design (see Figure 2.17). Delivering the project with a comprehensive BIM execution plan, workflows, information flows, and right resource allocations was the major responsibility of the INA BIM management team. Moreover, as the project was being delivered, continuous assessment through integrated project control and performance control were also conducted. It was aimed to have a transformative impact and lead to increases in productivity, efficiency and constructability of the project.

2.5 Conclusion

Throughout this chapter, key issues on airport design and construction, examples of on-going and/or BIM-enabled large-scale airport project deliveries, and also an overview of BIM-enabled project execution for INA were discussed. In light of the provided information on the design and construction of INA, one can say that the project has a lot to offer throughout its lifecycle. At the end of the completion of all phases, the project will come to life providing 76 million m^2 of airport with 6 runways, supporting 3,500 take-offs and landings per day, 200 million passengers a year, and access to 350 worldwide destinations. The project is one of the largest investments in modern Turkish history and is expected to create many employment opportunities for local people.

The new airport location moves airport traffic and support services away from a heavily congested portion of the city of Istanbul, and introduces a commercial economy to an area that is

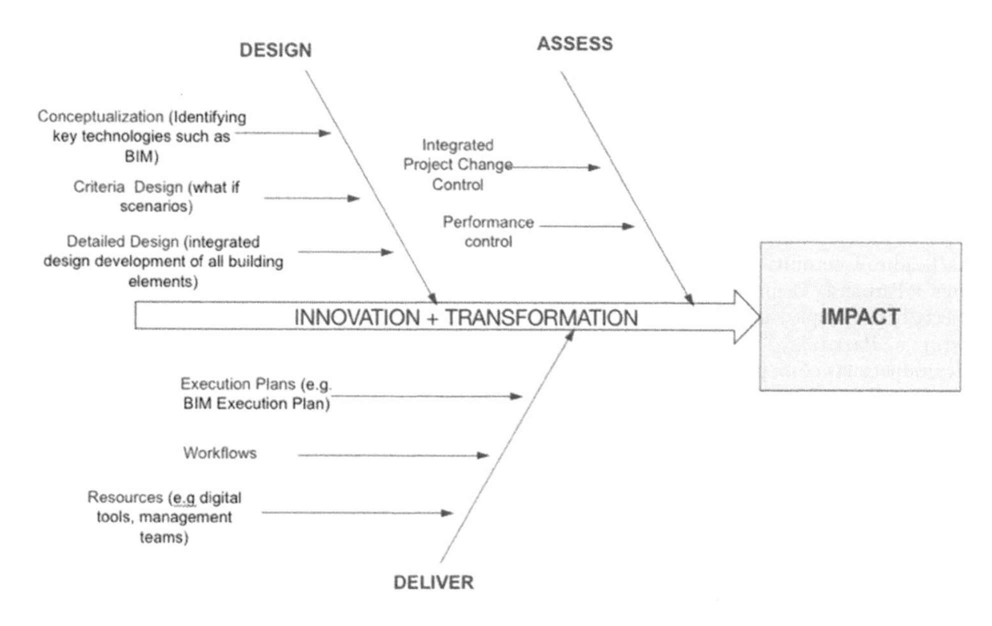

Figure 2.17 INA BIM management strategy with IPD mindset

currently subject to heavy mining. As a projection of global increase in air traffic, the trend of rising passenger and air traffic movements through Istanbul, and the desire to be recognized as a major hub for international air transport, parallel activities of design and construction centralizing BIM has been strategically adopted for the INA project.

In essence, the new airport aims to provide the following outcomes after the completion of the construction phase:

- New airport, new layout and development plan
- Available slots for new flights
- New regional planning (airport city, cargo city)
- Flights with very large aircraft
- Improved transit flight/passenger services
- New and rapid transportation systems

Lastly, the Istanbul New Airport project was selected as a finalist in the infrastructure category of the Autodesk 2016 AEC Excellence Awards. Also, at the 2016 World Architecture Festival, the project was awarded under the category of Future Projects – Infrastructure. The ATC tower stood out from among 370 projects to win this International Architecture Award.

Notes

1 IGA, "Construction." [Online]. Available: www.igairport.com/en/about-iga/construction. [Accessed: 23 March 2019].

2 IGA, "Construction." [Online]. Available: www.igairport.com/en/about-iga/construction. [Accessed: 23 March 2019].

3 Foster + Partners, "Kuwait International Airport." [Online]. Available: www.fosterandpartners.com/projects/kuwait-international-airport/#gallery. [Accessed: 23 March 2019].
4 The B1M, "Expanding Abu Dhabi International Airport with BIM." [Online]. Available: www.theb1m.com/video/expanding-abu-dhabi-international-airport-with-bim. [Accessed: 23 March 2019].
5 The B1M, "Expanding Abu Dhabi International Airport with BIM." [Online]. Available: www.theb1m.com/video/expanding-abu-dhabi-international-airport-with-bim. [Accessed: 23 March 2019].
6 Rogers Stirk Harbour + Partners, "Heathrow Terminal 5." [Online]. Available: www.rsh-p.com/projects/heathrow-terminal-5/. [Accessed: 23 March 2019].
7 Rogers Stirk Harbour + Partners, "Heathrow Terminal 5." [Online]. Available: www.rsh-p.com/projects/heathrow-terminal-5/. [Accessed: 23 March 2019].
8 Foster + Partners, "Beijing Capital International Airport." [Online]. Available: www.fosterandpartners.com/projects/beijing-capital-international-airport/. [Accessed: 23 March 2019].
9 Foster + Partners, "Beijing Capital International Airport." [Online]. Available: www.fosterandpartners.com/projects/beijing-capital-international-airport/. [Accessed: 23 March 2019].
10 Foster + Partners, "New International Airport Mexico City." [Online]. Available: www.fosterandpartners.com/projects/new-international-airport-mexico-city/. [Accessed: 23 March 2019].

Bibliography

Autodesk. (2017). *Foster + Partners and FR-EE take off with BIM.* Retrieved from: www.autodesk.com/solutions/bim/hub/aec-excellence-2017/infrastructure/first-place

Jones, S. A., & Bernstein, H. M. (2012) *The Business Value of BIM for Infrastructure: Addressing America's Infrastructure Challenges with Collaboration and Technology.* Bedford: McGraw-Hill Construction.

Reuters. (2019, January 3). *Building of new Mexico City airport suspended, but some works continue.* Today. Retrieved from: www.todayonline.com/world/construction-officially-suspended-new-mexico-city-airport-minister

The Vision 2050 Report (2011) IATA. https://www.iata.org/pressroom/facts_figures/documents/vision-2050.pdf

Uffelen, C. V. (2012) *Airport Architecture.* Berlin: Braun.

3 Airport Building Information Modelling

CONTENTS

3.1 Introduction

In this chapter, the concept of BIM strategy, planning for implementation, execution of adoption, control mechanism and planning for the future use of BIM are explained. BIM use for airport construction is a way forward for BIM implementation beyond buildings. Airport design and construction is highly complicated and incorporates a varying mix of infrastructure – including buildings, terminals, runways, passenger gates, car parks, railways and roads. INA is an extremely complex project while being one of the largest investments in modern Turkish history, where different international and local companies have been involved in its design and construction. Collaboration and coordination amongst them required a structured and holistic approach. A platform integrating all these related parties was therefore needed for on-time and on-budget completion of the project.

The level of complexity of BIM implementation increases as the scale of projects expands, so having comprehensive knowledge of ABIM processes is of utmost importance. The direct correlation between the complexity and demand for BIM implementation also adds value for better understanding of ABIM processes.

Not only easing airport project delivery, but also enhancing the business value of airport projects has led to an increase in BIM adoption. The global competitiveness of the AEC industry has led emerging markets, especially, to come up with "signature projects" including many mega-size airport projects, which require implementation of a high-level management system as satisfied by ABIM. Accordingly, ABIM is an integrated, iterative and self-assessed procedure, which boosts business value and highlights the project success.

From a holistic perspective, ABIM can be decomposed into determination of key performance indicators of the airport project; defining the tools and procedures for developing and executing the implementation of BIM at all stages, starting from design and proceeding with construction and operation; and utilizing and diffusing the digital transformation.

3.2 BIM strategy

The BIM lifecycle strategy spanned across design development to construction and final handover over a 30-month programme, which would be further extended to airport operations.

Figure 3.1 depicts how the INA BIM strategy encompassed all lifecycle phases of the project. The INA BIM has been responsible for continuous digital integration of the airport's systems – structural, architectural, mechanical, baggage handling systems, electrical, infrastructure, special airport systems, and information technologies – throughout the whole project lifecycle. Also, this BIM for the whole project lifecycle strategy led to an increase in the level of detail from LOD 200 to LOD 500.

Many parties and processes have been involved in the BIM implementation strategy for the INA project, as shown in Figure 3.2. BIM model creation started with design parties –which were responsible for structural, architectural, MEP, IT, SAS and other infrastructure systems – delivering their documents to the BIM department. Not only design documents, but also general specifications were considered while creating the BIM model.

As the BIM production team was generating the BIM model via integration of the design documents, the BIM management team strategically used the produced model for various purposes including design management, information management, quality control and assurance, resource management and performance management. To serve these purposes, the BIM management team conducted continuous systems coordination through clash resolution, integration of the construction baseline to develop the 4D BIM model, generation of bills of quantities (BoQ) for project control, and enablement of digital QA/QC and test and commissioning on site. Moreover, the BIM management team facilitated communication between all project parties by circulating the coordinated model to help everyone leverage the BIM model for the production of 2D/3D documents.

The BIM management team, as the representative of the client (INA), was the focal point of the BIM implementation strategy as it enabled digital project delivery by bringing together all

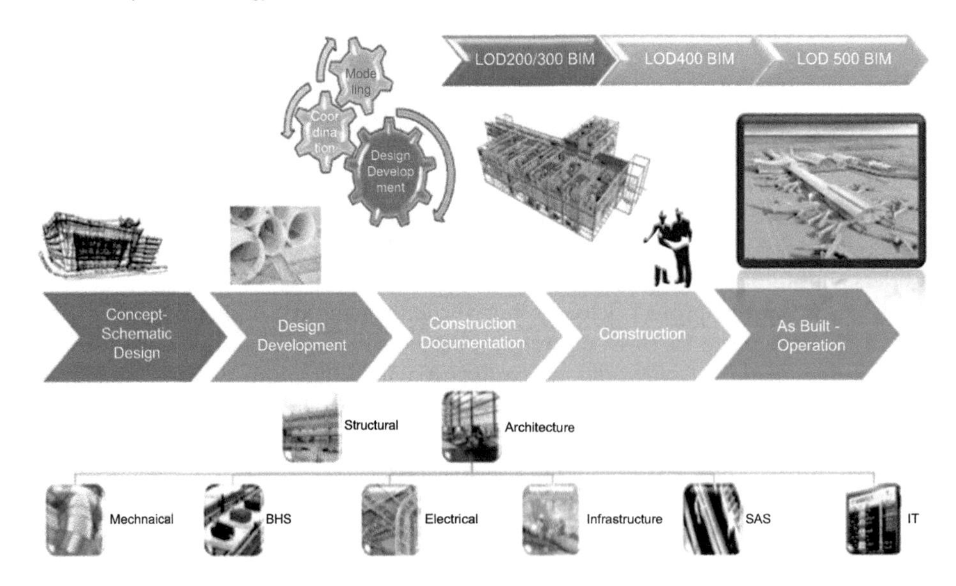

Figure 3.1 INA BIM strategy

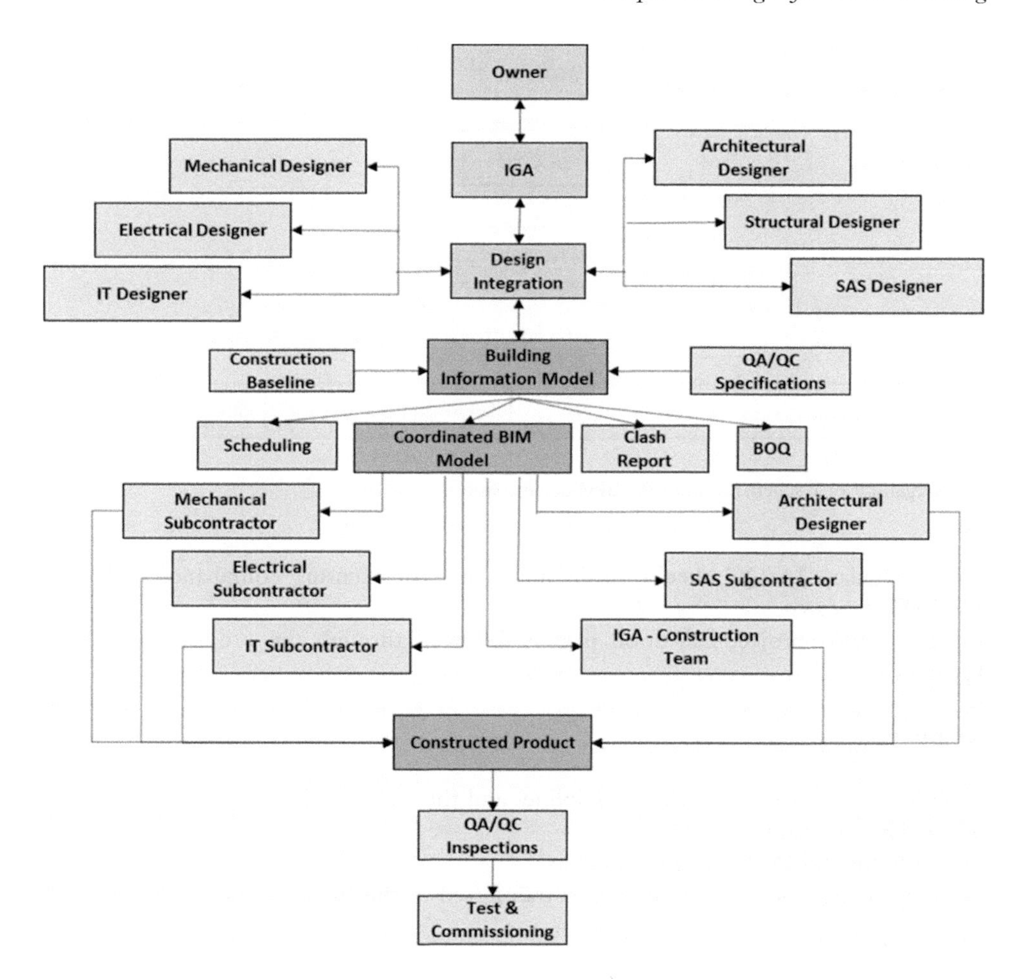

Figure 3.2 Relationships between the key project participants and BIM processes

key project stakeholders and technology in one platform. An organization schema of the INA BIM department is shown in Figure 3.3.

The responsibilities of each BIM management team member are listed below:

BIM director

- Create and execute the project's BIM strategy
- Review, monitor and approve overall BIM progress
- Manage and provide necessary support for BIM implementation in the overall project
- Report BIM delivery to the CEO and the Board

BIM manager

- Maintain the BIM execution plan
- Attend weekly BIM coordination meetings and BIM workshops

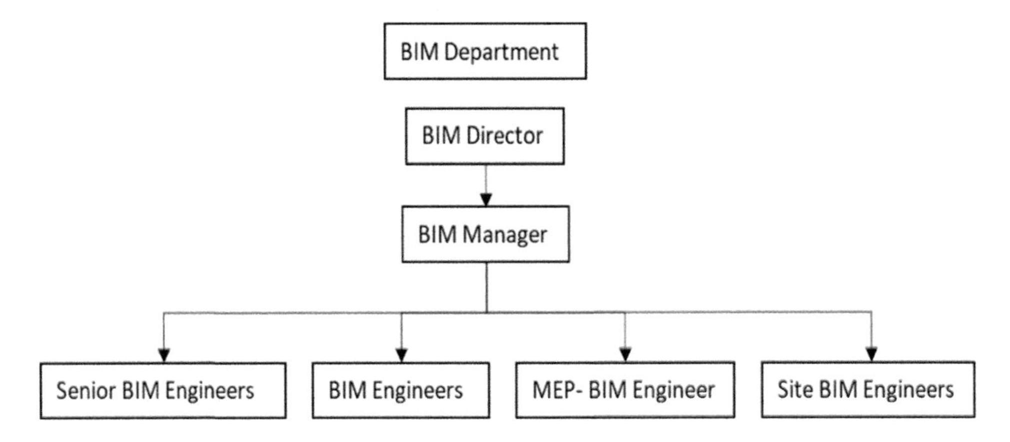

Figure 3.3 Organizational schema of the INA BIM department

- Perform regular QA/QC checks on discipline models to ensure compliance with the project's BIM standards
- Ensure the BIM project execution plan is followed through the project's duration on a daily basis

BIM engineers

- Establish communication between disciplines and the BIM production team
- Follow RFI and clash procedures
- Manage Vault and Buzzsaw environments
- Ensure up-to-date project information is transferred to the BIM production team

3.2.1 Workflows

BIM is executed across the project parties and leads manufacturing and the assembly of subcontractors on site. After initiation of ABIM implementation, BIM implementation workflows – such as the one shown in Figure 3.4 – were developed and put on action to plan how the information would be gathered, processed and delivered for the design and engineering processes of BIM.

The INA BIM workflow consisted of BIM integration, BIM production and BIM management. BIM integration included INA officials and subcontractors from all design disciplines and QA/QC. The first step was to gather the design information – models, documents, drawings and revisions – from the parties working on different design disciplines to generate the overall airport services/systems. The BIM management team filtered this received design information according to its consistency, relativity and also to the construction schedule, and then delivered the necessary information to the BIM production team for modelling.

Design integration, QA/QC checks and analyses, 2D design drawing inconsistencies and omitted information were gathered through raised Requests for Information (RFIs) from the BIM production team to the relevant disciplines. Major design issues were subject to RFIs. Thus RFI was the precursor before clash detection (clash detection and reporting format will be elaborated in detail in section 3.2.3). All RFIs and discipline responses were logged as

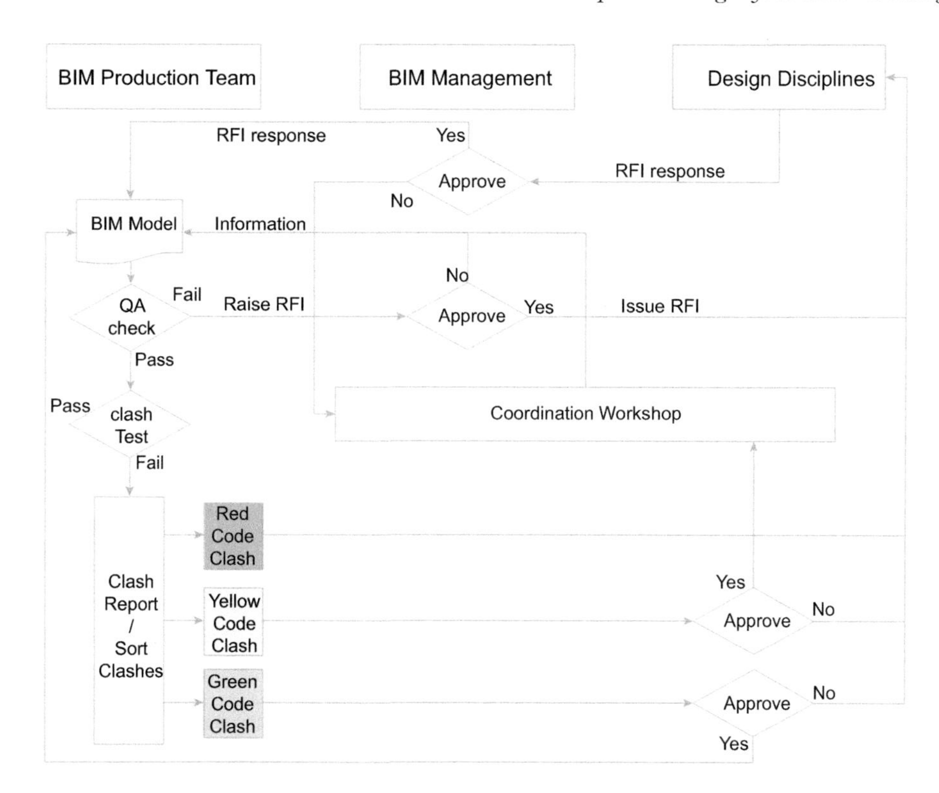

Figure 3.4 Decision model for BIM execution

open/closed by the BIM management team. Open RFIs were agenda items in BIM coordination workshops.

RFIs were critical for developing PD designs to IFC level. The BIM production team was responsible for integrating all design information into the 3D BIM model, which included but was not limited to 2D layouts, sections, elevations, and schedule and design specifications. The BIM production team worked under the administration of the BIM management team, and was responsible for regularly revising the models in line with the information provided and submitted to the asset and zone-wise models, both in Navisworks and Revit file format, due to the level of development according to the design and engineering phase respectively.

3.2.2 Meetings and workshops

BIM implementation at INA first took a role in sustaining the integration of all disciplines and related parties in a single virtual environment. Asset-wise meetings and workshops were scheduled to support BIM coordination among all project stakeholders starting from the design phase.

Workshops were held in the BIM room, which was equipped with a wide screen video wall and powerful workstations to establish an effective decision-making environment by enabling collaboration and communication through use of the BIM model. It is crucial to note that bringing people together in a common environment played an essential role in BIM success and the technological improvements it provided. The project individuals started to become familiar with both BIM products and the new way of collaboration enabled by BIM with the initialization of workshops via a common environment.

Creation of a common digital platform for communication and collaboration was achieved through meetings and workshops. As aforementioned, BIM coordination issues such as RFIs and clash resolutions were subjects for the BIM coordination workshops. BIM managers, INA design managers for relevant assets/disciplines and design consultants for relevant assets/disciplines attended the workshops in the BIM room. Hence, design for manufacturing and assembly (DFM) was achieved through streamlined communication.

BIM meeting and workshop types and the BIM room where all cross-coordination meeting and facilitated workshops took place are shown in Figures 3.5 and 3.6, respectively. The sequence and types of the meetings are further elaborated in Figure 3.7.

3.2.3 Monitoring and controlling

Processes were established to manage the subcontractors' deliverables. This process group was an important means of identifying variances in the subcontractors' deliverables from the BIM models. Continuous checking protocols were an integral part of the INA BIM department. The types of checks and their explanations are given in Figure 3.8.

BIM Meetings		
Meeting Name	**Interval**	**Attendees**
General BIM Coordination Meeting	Weekly	BIM Management, Design Managers From All Disciplines, Design Consultants From All Disciplines
Weekly BIM Production Meeting	Weekly	BIM Management, BIM Production
Daily BIM Production Meeting	Daily	BIM Management, BIM Production

BIM Coordination Workshops		
Workshop Name	**Interval**	**Attendees**
Terminal BIM Coordination Workshop	Twice a Week	Attendees: BIM Management, Design Managers From Relevant Parties, Design Consultants From Relevant Parties
Piers BIM Coordination Workshop	Twice a Week	Attendees: BIM Management, Design Managers From Relevant Parties, Design Consultants From Relevant Parties
Car Park BIM Coordination Workshop	Weekly	Attendees: BIM Management, Design Managers From Relevant Parties, Design Consultants From Relevant Parties
Airside-Sitewide BIM Coordination Workshop	Weekly	Attendees: BIM Management, Design Managers From Relevant Parties, Design Consultants From Relevant Parties

Figure 3.5 BIM meetings and BIM coordination workshop

Figure 3.6 BIM room

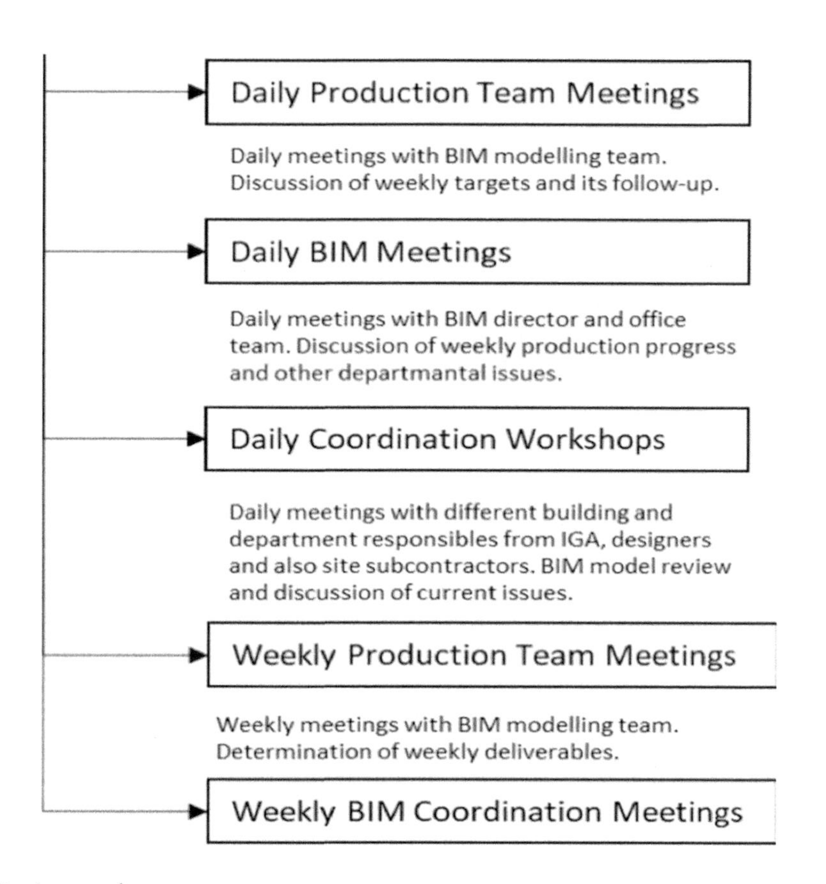

Figure 3.7 Meeting agendas

Type of Checking	Description/ Responsibilities	Frequency	Requirement of Checklist
Self-Check	Every production team member is responsible for this. They need to follow the checklist and review based on the LOD matrix	Regular Basis	Yes
Superior Check	Every BIM Production Lead is responsible for this. They need to check whether the BIM standards and model audits are maintained at production level	Weekly	Yes
Head Check	Every BIM Production Discipline Lead is responsible for this. The main part to check in this regard is design integration	Monthly	NA
Final Check	BIM Manager is responsible for this.	Prior to each Milestone Submission	Yes

Figure 3.8 Internal checking procedure of the INA BIM department

To control changes and recommend corrective actions to avoid future problems, the clash detection process was implemented through various monitoring and control mechanisms predetermined as part of the BIM strategy. Accessing the necessary information and resolving future operational problems related to the airport systems during the pre-commissioning, commissioning and maintenance stages was all possible with efficient usage of the BIM model. The participants working on the different design tools submitted their design information to the BIM management team, as agreed at the beginning of the project. By following a formal workflow, different data formats could all be gathered together.

Cloud-based data management tools were used to publish models, coordination documents and other related information via Buzzsaw and Vault applications. The data was gathered through the use of cloud-based information management tools. Defined individuals from each party were assigned these tools, and related permissions were given by the BIM team. Terminal architects were working on Revit, while all MEP-IT and concrete works were delivered as AutoCad files.

The airside surface and drainage models were provided in Civil 3D, and the baggage handling system was in AutoCad 3D. Even though all airport systems data was not in the same file format, the BIM production team managed to incorporate them into Revit. Thus, interoperability was not an issue as all the listed file formats could be exchanged through various import/export processes. One other key enabler for handling different types of file format was efficient cloud-based document management. Furthermore, after gathering all the data from different sources and different parties, the overall master model in Navisworks file format shown in Figure 3.9, which was exported from Revit, included all the latest project data to be used, reviewed and analysed during the workshops. It was kept updated with the latest information, which was discussed in weekly workshops.

INA chose Vault as the information exchange platform for the designers and consultants on the construction site. All exchange of files and information between the separate teams on the construction site was through Vault only. Buzzsaw was also chosen as the information exchange platform for remote designers and consultants. Similarly, all exchange of files and information between the separate remote teams was only through Buzzsaw. All the design data, drawings, models and reports were circulated through Vault and Buzzsaw in consultant/discipline-specific folders. The same folder structure was used in Vault and Buzzsaw to enable synchronization.

The BIM folder in Vault was continuously synchronized to Buzzsaw to ensure all parties benefited from up-to date information. All the information in Buzzsaw was then synchronized to a local server as a back-up procedure. At this point, BIM implementation in the project provided many opportunities by enabling straightforward access to all project data, including BIM models.

3.2.4 BIM model creation and coordination

The INA project was split into building zones to ease the management of planning, design and construction, which wouldn't be possible using only levels. Zones that were considered critical according to the construction schedule were reviewed with subcontractors in the BIM room during weekly workshops, and the required technical solutions were collectively decided by all the stakeholders in the BIM room and up-to-date design information maintained. BIM models were generated and revised periodically by gathering the latest design information from the project participants via the formal workflow and weekly workshops. This enabled timely clash detections and analysis for resolutions, before and during site installations. Delivery of these models to the construction site is explained in detail in Chapter 5.

Every two weeks, a clash report was generated by the BIM team. Discipline model elements were tested against each other according to the discipline clash check matrix. These issues were discussed in the weekly coordination workshops and models were coordinated and matched with the decisions in these meetings.

BIM model creation and its collaborative review provided a common visual environment for the systems forming the airport facility. Therefore, cross-disciplinary coordination and engineering

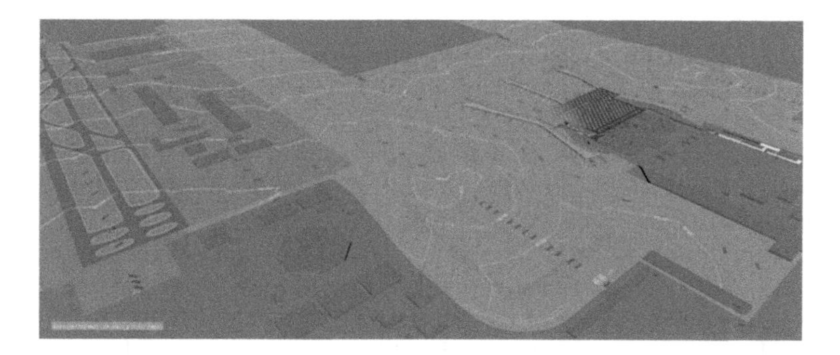

Figure 3.9 INA BIM master model image

decisions were accomplished under the supervision and guidance of the BIM department. This led to a quick transition from design to construction.

The BIM coordination process is based upon the construction schedule to lead production on site with accurate construction models. A design and construction schedule was developed to prioritize the zones that were nearing their site installation start date according to the schedule. Since the project was divided into high-level zones and assets, an Excel spreadsheet was created to follow upcoming construction zones. This Excel spreadsheet guided how coordination should be initiated to give an accurate construction model to the site for constructability purposes. This schedule was organized by combining the design drawing submission dates, BIM modelling timeframe, shop drawing production timeframe, coordinated shop drawing date and start date of construction according to the project baseline.

Design drawing submission dates were prepared asset-zone and level-wise by design managers for each service related to their discipline. BIM model coordination was facilitated, and seamless delivery to site was provided by the methodology of design, engineering for manufacturing and assembly. For example, a 1-week buffer was provided between the start date of construction and the finish date of coordinated shop drawing production. If there was less than 30 days left until shop drawing production, its timeframe was highlighted red, if 40 days were left it was yellow and if there were 60 days it was highlighted green; 30-40-60 periods were determined according to the observed mock-up zone coordination duration.

Navisworks was used in the project at all stages for coordination, clash detection and resolution and review purposes. The Navisworks file was used in workshops for coordination purposes. Model elements were managed through selection sets based on discipline key/critical areas and based on asset, discipline, level and zone.

The BIM management team was responsible for the distribution of the 3D Navisworks BIM model to the relevant parties. Navisworks models helped to manage selection sets and viewpoints to enable easier navigation for all users who had different 3D skill levels. Model elements could be selected according to discipline, zone or key areas.

Viewpoints were arranged to enable accurate viewing by zones or key areas. Clashes and issues were highlighted/clouded and saved as viewpoints. The Navisworks file was then circulated to all in NWD format weekly on Friday during the project. The BIM manager merged all the individual files into the master NWD file for the development purpose. The viewpoints related to RFIs were only added date-wise; and once RFIs were resolved, they were stored inside the Navisworks file.

After major coordination issues were resolved utilizing the RFI procedure, the model was in its basic coordination stage. At this stage, remaining issues were found using clash detection. Clashes were also identified according to their criticalities and practicability on site. Each discipline model was checked against other discipline models according to the discipline clash check matrix.

The clash results were reported to stakeholders in the INA clash reporting format. The idea was to remove all clashes and establish a clash-free coordinated model encapsulating the data required for quantification and operations compliance at later stages. For example, Figure 3.10 is the workflow created and adopted by INA for BIM MEP-IT coordination. This coordination process required close follow-up with the parties and quick resolutions. Weekly BIM workshops were therefore held to provide and maintain interdisciplinary coordination throughout the building/infrastructure assets in the airport construction.

Clash reports were produced to identify any issues in the latest design information collected in the BIM model. With this collaborative work, parties were able to visualize the latest status of design issues and were also aware of the effects of any possible revisions to the other parties. In this way, the design process was optimized by the parties. Any rework resulting from poor coordination and communication was also eliminated.

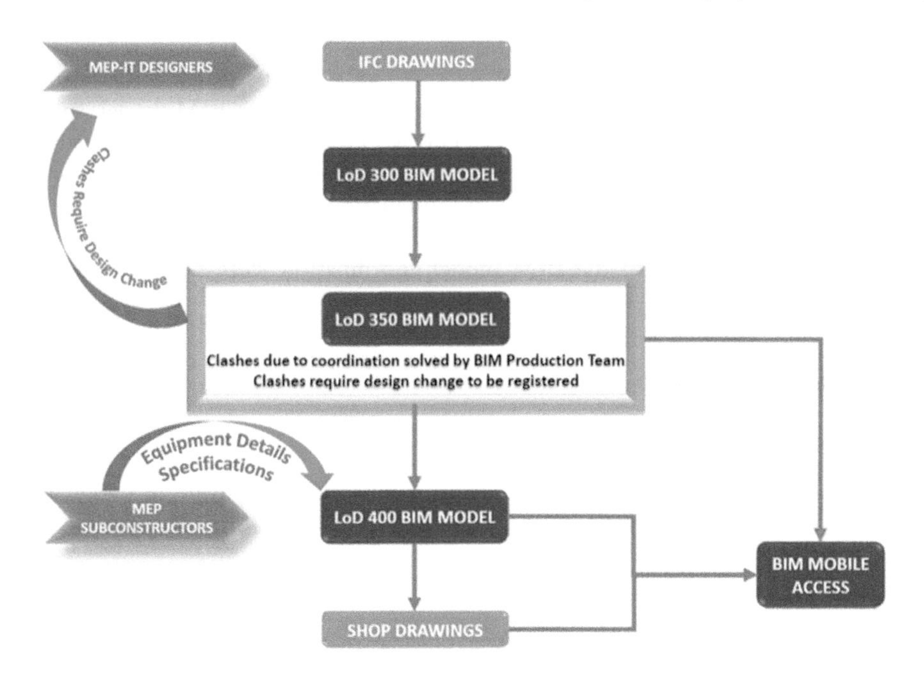

Figure 3.10 INA BIM MEP-IT coordination workflow

3.2.5 Clash reporting format

Every two weeks, a clash report was generated by the BIM production team. Discipline model elements were tested against the other discipline model elements to reveal the problems regarding design. Three levels of criticality for clashes were adopted in the project and these were treated in different ways.

1. **Critical severity:** Design change is required.
2. **Moderate severity:** Coordination is needed.
3. **Low severity:** BIM production team will resolve.

In general, most of the moderate severity clashes were resolved by the BIM production team upon approval of the discipline teams. Moderate severity clashes were to be resolved in the BIM coordination workshops. Critical severity clashes were the major clashes requiring design change. Major clashes were saved as viewpoints in the master Navisworks file submitted by the BIM production team. These viewpoints were used in the BIM coordination workshops and by disciplines individually to manage and solve issues.

Design coordination was realized within the disciplines of MEP-IT-BHS-architecture; and between MEP-IT-BHS-architecture and structure in the BIM environment. BIM provided a visualized environment for MEP-IT-BHS systems separately, and enabled the design teams to take their decisions under the supervision and guidance of the BIM department, which was the core coordination party of INA for the integrated design and construction delivery processes. BIM in INA was the integrative platform between all the related parties in the case of design and coordination.

3.2.6 2D drawings production procedure

The BIM model was part of the project deliverables, where relevant subcontractors were required to produce appropriate design products that were compatible with the BIM model, specified also in the contractual level as shown in Figure 3.11. This workflow was developed and put in use by INA to regulate the 2D drawing production procedure of subcontractors. MEP-IT subcontractors responsible for producing shop drawings for site installation were required to derive their 2D shop drawings from the BIM model, which was monitored by the BIM department. Drawings that were incompatible with the BIM model coordination were not accepted in a strict control process.

3.3 Terms and key concepts

3.3.1 Key concepts

BIM integration: The organization of officials and subcontractors from all design disciplines and QA/QC. The information and design deliverables (2D and/or 3D) from all disciplines

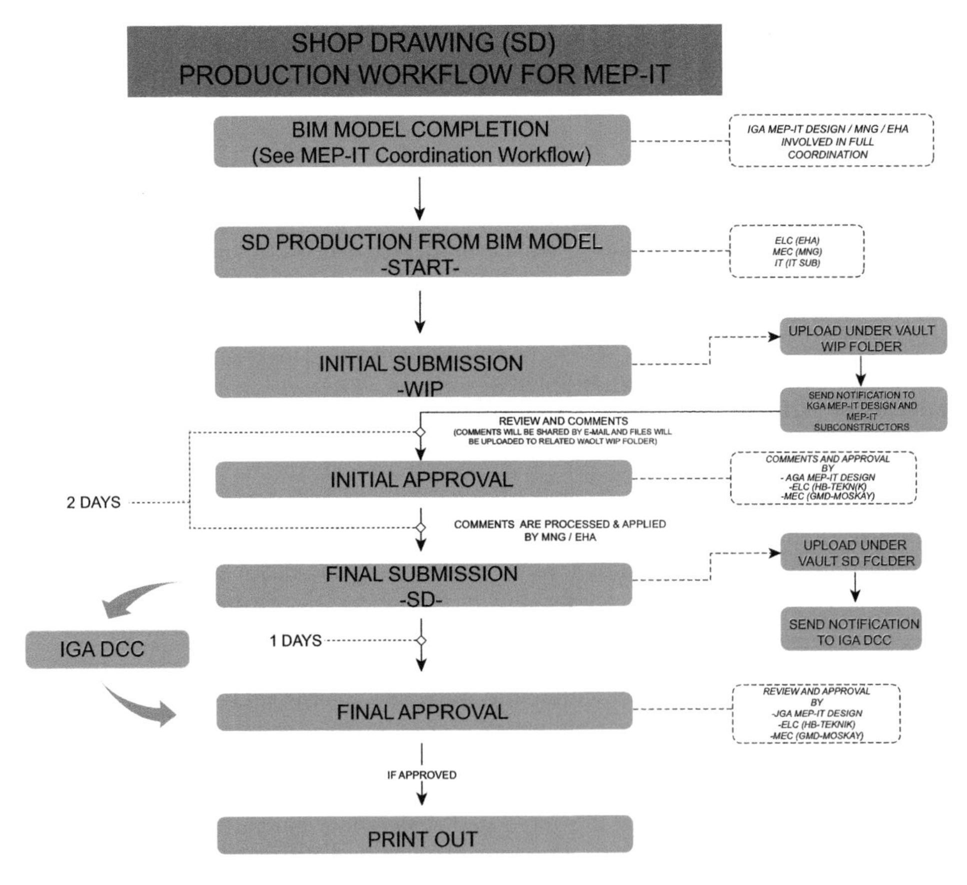

Figure 3.11 SD production workflow for MEP-IT

were submitted to Vault in 2-week intervals. Document submissions were logged by the project coordination team and reported to the BIM management team and submitting discipline officials after each submission.

BIM production: This is where all design information (which included, but was not limited to, 2D layouts, sections, elevations, schedule, design specifications etc.) was incorporated into the 3D BIM model. Vault submissions were synchronized to Buzzsaw continuously to ensure information on both environments was up to date. The BIM production team worked under the administration of the BIM management team and were responsible for BIM modelling, starting from LOD 200 (PD2), LOD 300 (IFC), LOD 400 (shop drawing) to LOD 500 (as-built). BoQ extraction from the BIM model, clash detection and reporting and 4D schedule simulation were performed regularly.

BIM management: The BIM management team created and executed the BIM execution plan. Project information from all disciplines was gathered through Vault and Buzzsaw environments. The BIM management team was also responsible for BIM-based project coordination, BIM model management and the integration of design, schedule, quantity and workshops where major RFIs and other project issues were solved.

BIM tools: These supported BIM processes at the project and organization levels and were generally categorized as authoring tools or audit and analysis tools.

Level of development (LOD): A description of the minimum dimensional, spatial, quantitative, qualitative, and other data which was included in a model element to support the authorized uses associated with such an LOD (AGC, AIA and NBIMS 2015).

3.3.2 Organizational structure

In this section, the organizational structure of the company is explained to demonstrate BIM execution through INA project participants. Project participants integrated with BIM included design firms, construction subcontractors and the BIM production team (see Figure 3.12). Whilst most of these firms were local companies, there were also companies from India, the UK and Holland which were integrated via BIM for the design and construction execution in the project master model. If the drawings were approved by both teams, they were shared with the site via tablets together with the integrated BIM models. If the drawings were rejected, they were sent back to the subcontractor for relevant corrections and updates.

3.3.2.1 BIM production team

It was discussed that the INA BIM management team would transfer the design data in 2D/3D forms to an outsourced BIM production team in India for 3D modelling and integration. There were daily follow-up meetings held between the BIM management team and the BIM production team to track the progress and deliverables. This outsourced team was selected due to their previous airport- and infrastructure-related BIM modelling experience. Although they had never worked on a project as large-scale as Istanbul New Airport before, they had completed similar types of project in the past.

There were, however, some issues that occurred regarding the deliverables requested by the BIM management team in terms of wrong or late delivery throughout the project process. The BIM management team experienced these issues when the work requested was not clearly explained to the production team. Since the staff of the production team were mostly focused on absolute modelling and didn't share the same concerns as the management team, the work given to them should have been described very clearly.

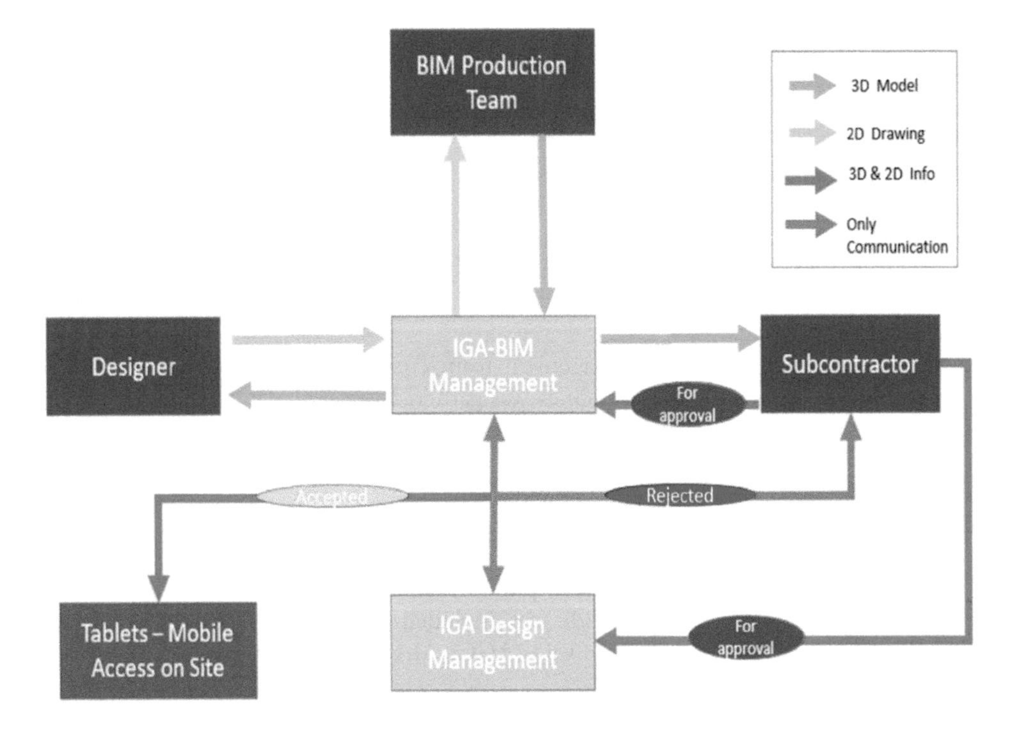

Figure 3.12 Data flow between project participants

Over time, continuous communications with the production team largely helped to avoid misunderstandings in the project. Since the production team were very open to the technological improvements, they were able to adopt new tools and integrate for mobile access. For each tool, they needed basic training on how to use and then implement them straightforwardly.

3.3.2.2 Designers

Design teams were responsible for submitting design information in 2D/3D formats to the BIM management team. For the architectural design, there were three local companies and each of them worked on different assets of the project: the terminal building, piers and car park. For example, the terminal design was carried out in Autodesk Revit and handed over to the BIM management team in 3D, whilst the pier and car park designs were produced in 2D.

For the structural design, there were local companies working on the steel and concrete structures of the project. The steel designers were designing in Tekla and passing design information in . IFC format for integration in Revit, while the concrete designers were working with 2D design tools. Mechanical, electrical and IT design were also conducted by local firms using 2D design tools. The baggage handling system designer was a Holland-based firm which used their 3D in-house design tool and transferred this in AutoCAD 3D format to the BIM management team for incorporation into the Revit model. For the runway and drainage designs, a UK-based design team used the Civil 3D tool and transferred their design as a CAD file.

The main difference between local and international designers was motivation towards innovation with BIM. Whilst the foreign BHS and runway designers were already using 3D tools and were familiar with the concept of BIM, many of the local firms had not worked in the BIM workflow before and avoided BIM adoption for a while. However, this cultural resistance was changed and mitigated during the weekly workshops by showing them the problems in their design, which could only be detected in the 3D BIM environment. They had the opportunity to monitor the status of their design as integrated with the adjacent services. Experiencing these opportunities with BIM provided the local companies with the confidence to adapt to BIM in the project.

3.3.2.3 Construction subcontractors

The construction subcontractors were all local firms and none of them had used BIM previously, except the BHS subcontractor. The BHS designer was also installing BHS services on site. They used their 3D model to manufacture BHS components for installation on site. They transferred their model to the BIM management team for interdisciplinary coordination. Owing to the integrated workflow, they were the most successful subcontractor in terms of timely completion despite the complexity of their design.

The remaining subcontractors received the coordinated Revit models from the BIM management team to complete their construction drawings. Even though they directly exported 2D sheets from the model, they could only use Revit models as reference and had to complete their drawings manually due to their lack of experience working in Revit. Although the companies were interested in working with BIM, there were not many people in Turkey trained in the area of BIM modelling at the time. That is why the major modelling job was outsourced to India.

Once the subcontractors had completed their drawings, they shared them with INA for approval. Their drawings were checked according to the coordinated models and approval was given accordingly. At the beginning, many inconsistencies were found in the subcontractor drawings. Due to the strict rejection of these drawings and subsequent affect on progress payments, the subcontractors had to use the BIM models. In order to use BIM models in Revit, they employed personnel with skills and understanding of the 3D BIM models.

3.4 Conclusion

To recap on this chapter, it is important to state that aligning the key concepts adopted in the INA project for airport BIM implementation with the organizational structure was crucial to achieve a scalable and agile ABIM platform for all project stakeholders. As demonstrated in this chapter, it is essential to understand project management processes, strategize BIM implementations and relocate resources accordingly.

4 Concurrent design and construction with BIM

CONTENTS

4.1 Introduction

In this chapter, BIM functions and strategies for the detailed design and construction phases are explained. In terms of preliminary engineering, BIM has provided a visualized environment for MEP-IT-BHS systems separately, and enabled the design teams to take decisions under supervision and guidance of the BIM department, which was the core coordination party for INA for design to construction delivery processes. BIM in INA was the integrative platform between all the related parties for design and coordination.

4.2 Lean perspective for concurrent engineering

Lean construction is a method to design production systems. These systems are mainly targeted at reducing the waste of materials, time and effort to gain the maximum possible amount of productivity and value in construction working processes (Aziz and Hafez, 2013).

Although lean construction and BIM have emerged as separate domains in construction IT research, there are substantial synergies between them that have led to the combined use and implementation of lean and BIM methods (Dave et al., 2013). On the other hand, their parallel adoption in state-of-the-art construction practice is a potential source of confusion when assessing their impacts and effectiveness. Therefore, a rigorous and systematic assessment based on real facts and figures is also necessary, since the construction industry has much to gain by applying lessons from efficient practices. For example, in manufacturing industries, product lifecycle management (PLM) systems have been refined for decades, enabling the extended collaboration principles to systematically reduce cost, improve sustainability and maximize value. Hence, BIM and lean construction use is also required for extended collaboration and for an end-to-end cooperative process to reap lean efficiency gains and reduce waste for designers, builders and operators.

Lean design and construction methods share the same objectives as lean production, e.g. cycle time reduction, elimination of waste and variability reduction. Continuous improvement, pull production control and continuous flow are the end goals for the implementation of lean construction (Ballard, 2008). The most valuable advantages of lean construction are the workflow reliability and value streams as there are big gaps in the traditional construction practices, causing waste and loss in value. The construction industry is facing demands to (i) increase productivity, efficiency, infrastructure value, quality and sustainability; and (ii) reduce lifecycle costs, lead times and duplications via effective collaboration and communication of stakeholders in construction projects (Nour, 2007).

BIM and lean construction are two different initiatives to improve the construction process with distinct communities and interests. BIM and lean construction have the same working principles with regard to construction practices. Both have started to diffuse into construction practice with an accelerating speed of process, and an integrated synergy with BIM and lean practices has been a well accepted theory by academicians (Koskela et al., 2002).

The lean and BIM workflows span the entire lifecycle of a project. Figure 4.1 shows the main applications executed in an integrated lean and BIM project. From this point of view, there are increasing requirements to make BIM leaders and users aware of lean principles, methods and tools, and lean leaders and implementers aware of BIM. It is important to use the same language related to BIM capabilities (Koskela et al., 2002). The common ground that exists between lean design and construction and BIM can be summarized with the following features:

- Eliminate waste – minimize planning, features, metrics and documentation, which would be possible with BIM using structural clash tests and producing design alternatives to select the most suitable design and performance simulation for energy efficient solutions.
- Amplify learning – positive iteration and client involvement, which would be possible with BIM by using visualization solutions that ensure clear understanding about design, collaborative analysis for the best results, and improving communication between the client and suppliers by using 3D models and walkthroughs
- Decide as late as possible – options-based approach, which would be possible with BIM by using visualization of the workflow to check for process conflicts (teams and tasks) and provision of accurate and complete information to prefabrication and shop drawings for construction execution

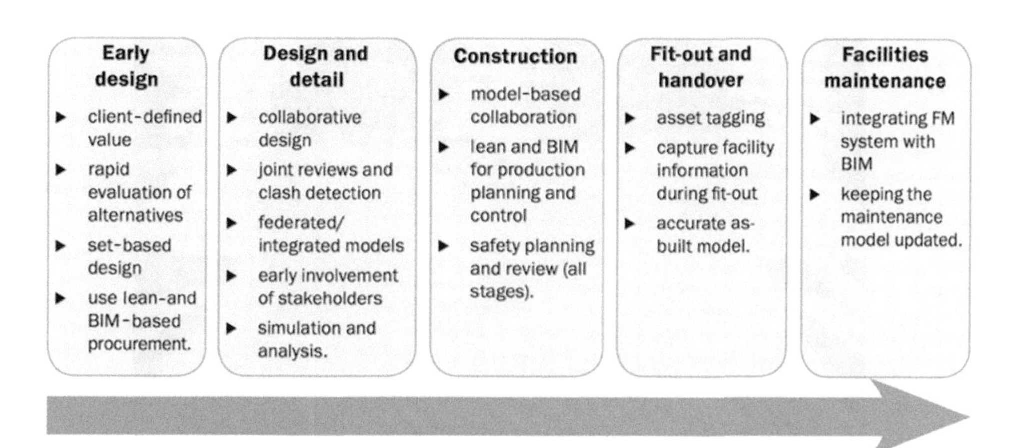

Figure 4.1 Lean and BIM workflow – functions (Dave et al., 2013)

- Deliver as fast as possible – rapid value flow and iteration of needs, which would be again be possible with BIM via automated generation of changes and materials schedules and quantities, and collaborative workflow and information sharing
- Empower the team – facilitate team commitment and rapid feedback, which is possible with BIM via accurate and complete information sharing throughout the building lifecycle
- Build integrity in – conceptual and perceived usefulness over time via BIM through making detailed schedules of task and material delivery times in an integrated manner
- See the whole – avoid sub-optimization via collaboration and concurrent working on the project by different stakeholders or teams.

In the next section, the case study INA construction project is briefly introduced, where these lean and BIM synergies are examined.

4.3 Digital design with BIM

BIM implementation during design and engineering in the INA project created a common virtual environment for all the parties involved in the process. This was enabled by data exchange procedures via cloud-based data management tools, and the development of BIM models with necessary engineering decisions by integrating different forms of design information.

To utilize and sustain these practices, participants gathered periodically and procedures were followed as determined and published in the BIM execution plan. This practical implementation improved the quality of the design product, which would finally be applied on site since BIM provides the latest design and engineering information with a clash-free prototype. The integrated BIM model shown in Figure 4.2 illustrates the complexity of the structure and how BIM aided clash detection in the INA project.

Since interdisciplinary coordination was performed using the BIM model, design errors were detected and addressed to the relevant parties prior to construction execution on site. Any revisions were updated directly in the BIM model and its consequences were monitored closely. These revisions were also shared with all the related parties. Consequently, an accurate design

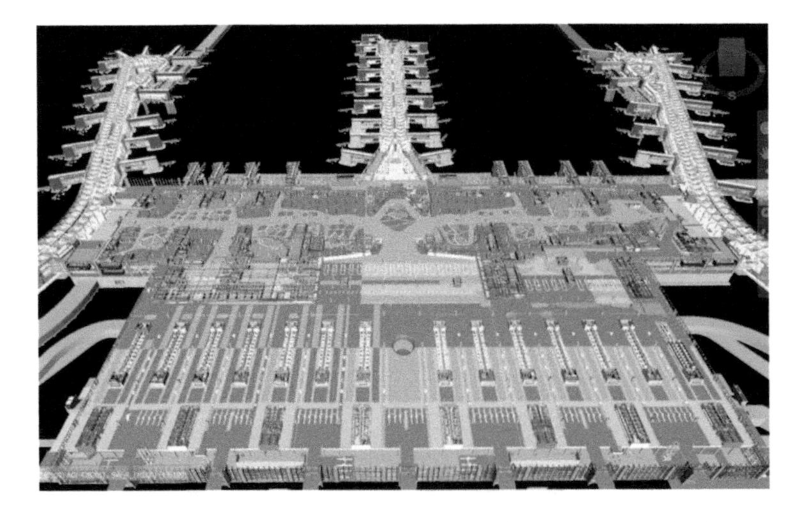

Figure 4.2 The integrated BIM model (MEP+structure+architecture) for design coordination

product was generated for high-quality and fast production on site by keeping design parties up-to-date and leading the construction team to zero-defect and effective installation. Since stakeholders and subcontractors were also involved in the processes, they worked actively during the creation of the final product to be installed on site.

In the INA project, more than 600,000 clashes were solved, saving time and money and preventing rework on site. Any unexpected claims, time extensions and cost overruns were also prevented. Figure 4.3 also shows a perspective image from the integrated BIM model used for clash analysis.

The priori impact and lean efficiency gains from these 600,000 clashes are tabulated in Figure 4.4. The clashes are categorized as major, medium and minor clashes. Each of the major clashes requires 45 days of fixation with 10 workers on average if they are encountered on site in a traditional practice, rather than the BIM-based practice, whereas medium-size clashes need 21 days with 5 workers on average and minor clashes require 2 days with 2 workers on average. In a design and construction scenario based on a traditional practice, expecting the encounter of those 600,000 clashes would normally cost around an extra 2.5 billion Euros and more than 10 years of overtime in the project.

In Figure 4.4, cost and time savings are computed with the normalized values. In other words, dependency between clashes are also taken into consideration: If a clash is resolved, some other clashes are also subsequently resolved due to relevance and correlation between them. For example, when a major clash is explored and fixed in the combined BIM model, there are possibly 4 or 5 other clashes are also resolved, which is therefore normalized with 0.25, while medium clashes are normalized with 0.33 and minor clashes with 0.5. As a result, the normalized clash numbers are calculated accordingly. In the normalized clash numbers, a clash and its dependent clashes are considered as one clash. This leads us to the total of normalized clashes, which is 238,635 clashes altogether so far.

In terms of lean principles, this means a time saving of 16.442.036 days and a cost saving of 835.389.260 Euros in the project. This is approximately 10% of the total budget for Phase 1 of the INA project. This amount would be around £2.5bn if the raw clash numbers (601,918) are considered, which would be the case of time delays and financial burden if the project had used traditional practices. These numbers already prove the synergies between BIM and lean principles in terms of eliminating waste and generating values.

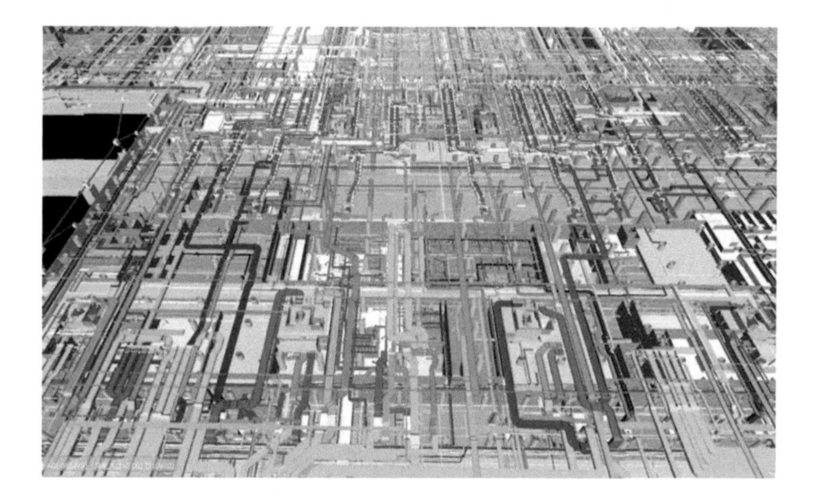

Figure 4.3 A view from the integrated BIM model used for clash detection

TERMINAL	Number of Clashes	Clash depency Coefficient	Number of Clashes with dependencies	Per Clash /Man*day	Total Savings Time (Day)	Labor Cost (35€/Per Day)	Reconstruction (Avg: €500)	Total Cost Saving
BHS vs MEP Clashed	52.305	0,25	13.076	*45 days *10 men	5.884.200	€205.947.000,00	Negligible	€205.947.000,00
MEP vs MEP Clashed	240.176	0,33	79.258	*21 days *5 men	8.322.090	€291.273.150,00	Negligible	€291.273.150,00
MEP vs ARCH & STRC	76.275	0,5	38.138	*2 days *2 men	152.552	€5.339.320,00	€76.276.000,00	€81.615.320,00
BHS vs ARCH & STRC	96.750	0,5	48.375	*2 days *2 men	193.500	€6.772.500,00	€96.750.000,00	€103.522.500,00
PIERS								
MEP vs MEP Clashes	49.521	0,33	16.342	*21 days *5 men	1.715.910	€60.056.850,00	Negligble	€60.056.850,00
MEP vs ARC & STRC	86.891	0,5	43.446	*2 days *2 men	173.784	€6.082.440,00	€86.892.000,00	€92.974.440,00
TOTAL	601.918		238.635		16.442.036 man-days	€575.471.260,00	€259.918.00,00	€835.389.260,00

Figure 4.4 Clash detection and analysis and related time and cost savings achieved

Notes: BHS = baggage handling system; MEP = mechanical, electrical and plumbing; ARCH = architecture; STRC = structure.

INA regulated the 2D drawing production procedure of subcontractors. MEP-IT subcontractors responsible for producing shop drawings for site installation were required to derive their 2D shop drawings from a BIM model, where the process was monitored by the BIM department and relevant contractor designers and the drawing approval process was based on this approach. Drawings that were incompatible with BIM model coordination were rejected in a strict control process.

4.4 Digital construction with BIM

Building Information Modelling played a crucial role in facilitating the development of the INA project. The BIM department in INA managed the coordination between the design and engineering firms including structure, architecture, MEP and BHS. BIM had a strategic role in executing engineering and design to accelerate efficiency in design and construction, which was a key driver for delivering the project on time and even ahead of schedule. Figure 4.5 shows the supply chain integration amongst the design and construction firms involved in the INA project via the BIM department. The BIM department ensured that the required project information was available to whoever needed it, whenever it was needed and wherever it was needed. Publishing and exchanging the correct data was crucial. Cloud-based data management tools were used to publish models, coordination documents and related information.

As shown in Figure 4.5, that coordination process required close follow-up with the parties and quick resolutions; weekly BIM workshops were held in order to provide and maintain interdisciplinary coordination throughout all of the building/infrastructure assets in the airport construction. Clash reports were produced to identify any issues in the latest design information collected in BIM. With this collaborative work, parties were able to visualize the latest status of design issues and also be aware of the effects of any possible revisions on the other parties. In

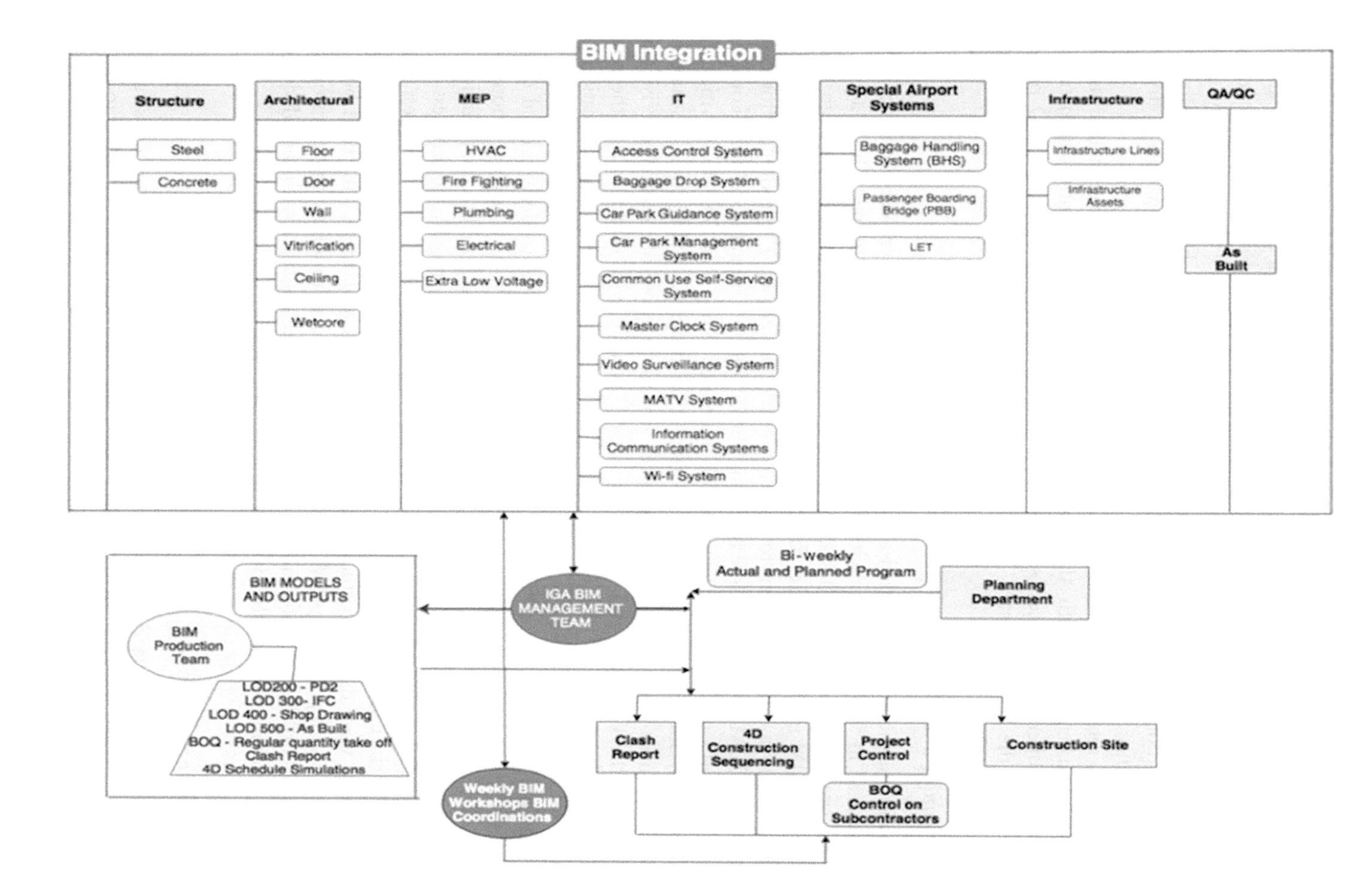

Figure 4.5 INA integrated construction environment via BIM workflow

this way, the design process was optimized by the parties, and any rework resulting from poor coordination and communication was eliminated.

The process of designing, integrating and documenting the design and construction information by developing an integrated BIM model evolved through the design and construction stages of INA for the BIM-based project delivery. Automated data processing via an integrated environment across the project stages also reflects a lean design and construction practice.

The resulting clash-free – ready to manufacture – BIM model with necessary engineering decisions considering all disciplines was shared to the site directly by mobile applications. This led to faster and more accurate application of the coordinated design without any delay and with fewer problems on site.

However, BIM integration in INA was not limited to design coordination but also aimed to improve project management with 4D construction sequencing, 5D quantity estimating and also construction site supervision processes. The sections that follow expand on the 4D planning and cost estimation in the INA project.

4.4.1 4D planning and 5D cost management

4D construction sequencing integrated the baseline schedule of the project with the BIM model and was discussed in the BIM room with the construction teams in terms of both actual progress and possible site installation clashes for critical activities. With the up-to-date design information collected in the BIM model, quantities for each discipline were extracted in the format agreed with the technical office and used by them to control the subcontractors. These office efforts were also controlled by the BIM engineers on site to ensure the decisions were applied during construction.

An enormous 4D model was developed that integrated more than 30,000 activities to enable daily and monthly monitoring of the INA construction progress, giving the construction teams dynamic control over progress in the INA project. For the purpose of executive decision-making, 4D snapshots of the key assets and zones were shared weekly with directors and the responsible bodies. In addition, the 4D BIM model was used for coordination and communication with the related stakeholders, who also attended the BIM room meetings.

The main advantage of using BIM for quantities management is to derive actual quantities of items used in the project to eliminate any excess amount and verify the estimated amount. Furthermore, a bill of quantities can be generated from the BIM model and issued to the technical offices of each discipline. Figure 4.6 shows the workflow of 4D/5D integration with site progress reporting.

The workflow shown in Figure 4.6 explains how BIM-based project progress monitoring applied in INA. Since planning, quantities and QA/QC were directly carried with BIM and the progress payments were done accordingly, subcontractors had to comply with BIM both for their work on site and also reporting to INA. This meant that they had to follow the milestones given in the baseline, install on site by conforming to QA/QC standards together with the BIM model, and report the quantities in line with the BOQs extracted from the BIM model.

4.5 Lean efficiency gains

The lean efficiency gains via design and engineering, BIM procedures and construction management in INA were realized through BIM utilization. With the use of a 4D model, progress monitoring was simulated and observed by the management team. Any possible delays were addressed and prevented before they occurred, while 5D integration provided effective quantities management. The latest and most accurate information was derived from the BIM model.

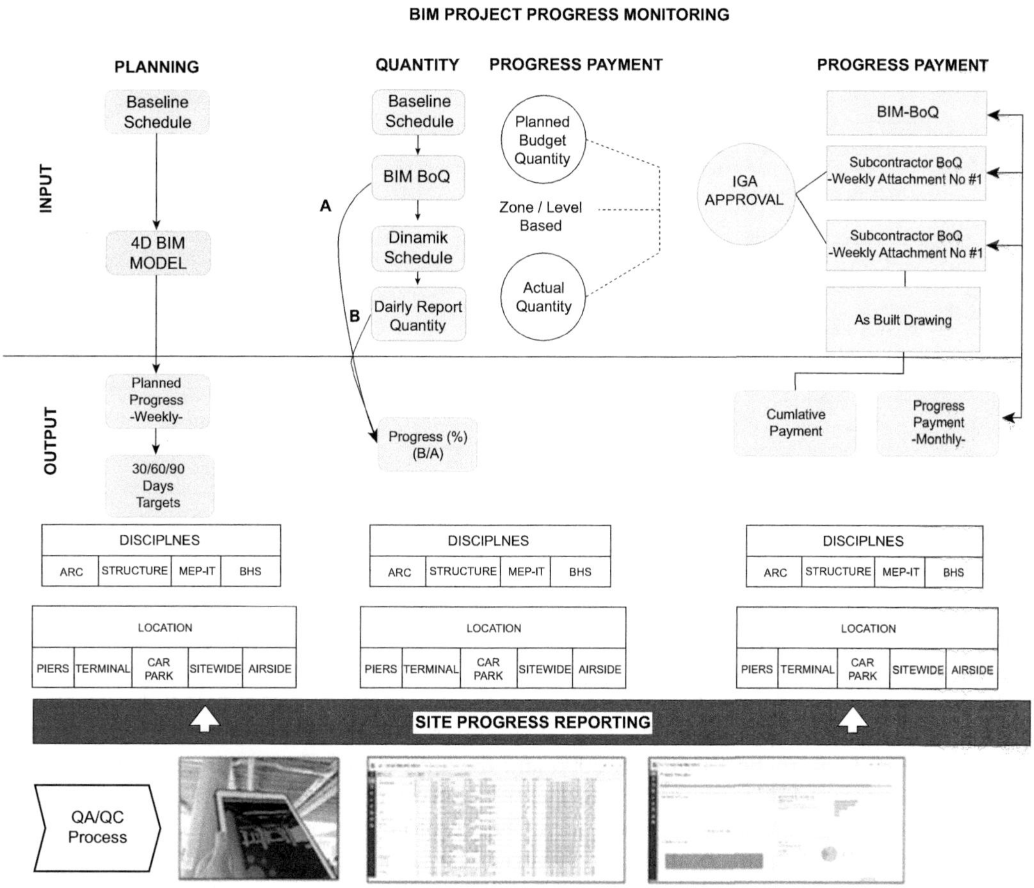

Figure 4.6 BIM-based project progress monitoring

By delivering a constructible BIM model to site, the construction team had all the necessary and applicable information, which led to faster and more accurate installation on site. Since all the issues were resolved and the latest information was shared with the site, rework possibilities were minimized. After installations were fulfilled on site, they were also inspected with BIM-based QA/QC procedures and the records were stored digitally in a cloud system. Digital QA/QC procedure reduced unnecessary paperwork and led to an effective workflow automatized via BIM.

While storing QA/QC records digitally, any high-level reporting, which could be generated periodically or spontaneously in various forms to show progress, completeness, trend or details of the defects, was utilized and distributed to the responsible departments. This drew a clear picture for spotting errors and made it possible to take preventative actions to facilitate zero-error production. Since everyday site installation practices were handled and improved with BIM practices, error reduction was made possible, which led to the lean production.

In this way, it was ensured that the test and commissioning phase of the INA project was initiated with zero defects. With the use of BIM models on site, any as-built and facility information was captured on site and asset tagging was performed to complete passage to facilities management. The lean efficiency gains can be better appreciated by comparing the INA case

Project	Istanbul Grand Airport	Berlin Brandenburg Airport
Total Size	76.5 million m2	14,7 million m2
Indoor Area	3.5 million m2 (phase 1 only)	1.8 million m2
Number of Passenger	200.000.000	24.000.000
Number of Employee	30.000	1.500
Budget	€10,25 bn	$2 bn
Planing Start-Finish Time	2014 - 2019	2006-2011
Construction Method	BIM	Traditional Method
Save/Over Cost	+ €2,5bn	- €6bn
Save/Over Time	On time	2019

Figure 4.7 Comparison of Berlin Brandenburg Airport and Istanbul New Airport projects

with the Berlin Brandenburg Airport case, where traditional methods and processes were used. Figure 4.7 tabulates the key findings between the two airport construction cases.

Berlin Brandenburg Airport was expected to become the third busiest airport in Germany, with a projected 45 million passengers annually. The airport's feasibility and preplanning phase took about 15 years. Construction started in 2006 and the airport took 5 years to be built. The target opening date was October 30, 2011 (BER, 2011). However, at the time of writing this book, Berlin Brandenburg Airport has yet to open. The project was originally set at a total cost of 2 billion Euros (2.1 billion US dollars). However, the latest estimate of the total project cost is 7.9 billion Euros, almost 50% above the approved budget of 6 billion Euros (*The Economist*, 2017).

The company that runs the airport, which is owned by the city of Berlin, Brandenburg State and the federal government, spends 17 million Euros each month in maintenance for the empty terminal building, while forgoing some 13 million Euros in rental income. The airport is now scheduled to begin operating from 2020 and onwards (BER, 2011).

The Berlin Brandenburg Airport flight paths and sound protection zones were incorrectly calculated. Reports indicate that 66,500 defects were found; 34,000 are described as "significant" and 5,845 as "critical". Among the construction faults are: the faulty fire protection system; the ventilation system which still does not work; car parks that began to crumble weeks after they had been completed; missing check-in counters and luggage conveyor belts; faulty fire-safety walls between the airport and the railway station that serves it; and pipes and cables so ill-fitted as to be useless. Unlike other major projects, the architects had planned to funnel smoke – which usually rises – underneath the airport's halls. Significant reconstructions were planned to remedy the disastrous situation (Ondruskova, 2014).

In the case of poor management of critical congested areas in such a complex infrastructure project, the following outcomes can be expected to occur:

- Excessive capital costs
- Excessive construction time

- Lack of predictability
- Unacceptable levels of defects
- Lack of productivity
- Lack of facility management process
- Lack of sustainability

On the other hand, the greatest advantage that the BIM methodology has given the INA project in comparison to a traditional approach, apart from greater control of the project and all its constituent parts, is release from the tedious work of information production and repetition of processes. By automating production and allowing more time to be dedicated to the design phase, BIM results in projects of higher quality, with less errors, deviations and problems in their construction and, finally, tremendous time and cost savings as shown in Figures 4.4 and 4.7. The following outcomes are achieved via ABIM:

- Capital costs as estimated
- Construction time as scheduled
- High and correct predictability
- Hardly any defects on site
- High productivity
- Sustainable facility management process

It is strongly believed that the strategic BIM use in the INA project, which is more complicated and larger than the Berlin Airport construction, helped save the project and enabled substantial lean efficiency gains including time and cost. Furthermore, as a result of the timely completion, income generation from the operational management of the airport assets will be possible. Considering all these gains, benefits and value generation are substantially high.

4.6 Conclusion

Having explored the INA BIM team's strategy for leveraging the collaboration of key principles – lean construction and BIM – to increase efficiency in such a mega project, one can say that creating and diffusing a sustainable synergy for concurrent design and construction was made possible by gathering people, the right technology, and integrating key principles in the same platform. Lean principles were among these, and played a crucial role in the process innovation and evolvement that BIM introduced to the design and construction of the project.

Concurrent design and construction was made possible with BIM, enabling digitization both in virtual coordination and manufacturing on site. Furthermore, not only substantial time savings, but also facilitated communication between project stakeholders accelerated the value creation. Overall, making Phase 1 of the INA mega project executable in such a short timeframe of 3 years has been achieved by the efficient synergies between full BIM implementation and decisive principles like lean construction.

References

Aziz, R. and Hafez, S. (2013) Applying lean thinking in construction and performance improvement, *Alexandria Engineering Journal*, Vol. 52, No. 4, December, pp. 679–695. doi: 10.1016/j.aej.2013.04.008.

Ballard, G. (2008) Lean project delivery system: an update, *Lean Construction Journal*, No. 2008, pp. 1–19, available at www.leanconstruction.org/learning/publications/lean-construction-journal/lcj-back-issues/2008-issu.

BER (2011) *Berlin Airport Annual Report*. Berlin: Berlin Brandenburg Airport. Available at www.berlin-airport.de.

Dave, B., Koskela, L., Kiviniemi, A., Tzortzopoulos, P. and Owen, R. (2013) *Implementing Lean in Construction: Lean Construction and BIM*. London: CIRIA.

The Economist. (2017) Why Berlin's new airport keeps missing its opening date, *The Economist*, Berlin, available at: www.economist.com/blogs/economist-explains/2017/01/economist-explains-18.

Koskela, L., Huovila, P. and Leinonen, J. (2002) Design management in building construction: from theory to practice, *Journal of Construction Research*, Vol. 3, No. 1, pp. 1–16. doi: 10.1142/S1609945102000035.

Nour, M. (2007) Manipulating IFC sub-models in collaborative teamwork environments, CIB78: 2007 Series, The University of Ljubljana, available at: http://itc.scix.net.

Ondruskova, I. (2014) *Berlin Airport: The Five Biggest Mistakes*. Berlin: DW. Available at: http://p.dw.com/p/1CR8S (accessed December 17, 2017).

5 Mobile BIM for the airport construction

CONTENTS

5.1 Introduction

Construction management is the act of overall planning, coordination, organizing, overseeing and control of the tasks involved in a construction project from inception to completion, focused on client requirements to produce a functional, efficient and financially viable project that will be completed on time within budgeted costs and the required quality standards. The INA BIM department has managed the execution of the construction project through the planning, design and construction phases by considering the quality, cost, time and scope.

During the design and engineering phase in the INA project, BIM has provided interdisciplinary coordination via BIM model clash detection and daily workshops held in the BIM room. Since all of the relevant parties including subcontractors were involved in the process, 2D shop drawings produced for site installation purposes were also compatible with the coordinated BIM model. Apart from design coordination, the construction schedule was integrated with the BIM model to monitor progress and take necessary decisions to avoid any possible delays.

BIM-based quantity take-off was performed periodically and shared with the purchasing and contractor technical offices, which were controlling subcontractor installation quantity and progress payments according to these installations. Whilst the design team was progressing with the BIM implementation process and adoption, these office efforts all needed to be extended to the construction site to maximize BIM benefits. With the improved formal workflows enabled by technology and cumulative BIM efforts, the aim was to increase engineering quality, reduce unnecessary rework and make cost and project schedule savings.

5.2 Mobile platforms

Cloud-based collaboration tools Autodesk BIM 360 Field and BIM 360 Glue were selected to transfer the digital BIM data from office to site, and lead all the processes given in Figure 3.12 because of their user-friendly features and compatibility with the common tools used in the design phase. The aim was to keep the site updated with the latest and the most accurate design/construction data in the form of BIM models, shop drawings, MSs (method statements) and MAFs (material approval forms), whilst automatically monitoring and improving quality management workflows.

To run these applications on site, organization schemes of all the departments within the company were studied to determine the total number of tablets to be purchased. The required number of tablet users was discussed with the department heads and the final number was settled at 150 for the 6 departments: MEP, IT, BHS, SAS, superstructure, infrastructure and QA/QC.

As shown in Figure 5.1, site engineers were equipped with iPads on which only selected and approved tools were installed by the BIM department. Access to internet browsers or any other applications selected by BIM was restricted by the company's IT department via a mobile device management system. These arrangements ensured that the site personnel were focused on only work-related activities while using tablets and also ensured privacy of project data by restricting data share with any other device.

Site engineers and QA/QC personnel started to get familiarized with the new tools, workflows and communication ways via trainings given by BIM engineers, who formed and optimized the tools' structure for effective use according to the project's characteristics. Because of the size of the INA project, management issues on a daily/weekly basis were foreseen as a challenge. For this purpose, weekly sessions were set with all the users to upskill them in using the system. These sessions were not considered as start-finish processes, they continued periodically across all departments to educate newcomers and also keep current users updated with the features and workflows regarding the tools in use.

A BIM mobile access guide was prepared and distributed by the BIM department to create an easily accessible document for anyone involved in the project, as shown in Figure 5.2. Standards were also set for common processes to be followed by all personnel.

Figure 5.1 Site engineers using tablets on site

Figure 5.2 INA BIM mobile access guide

The aim was to steer the site installation progress while making it cost effective and high quality with minimized rework using BIM mobile access. Glue and Field tools were functional on site with the following contexts: coordinated BIM models; approved 2D shop drawings; QA/QC documentation including MAFs and MSs; QA/QC processes; inspection forms, inspection checklists and site observations; performance tracking; and crew and jobsite planning.

5.2.1 Mobile tool set-up

In the cloud, selected tablet users, as explained in the previous section, were identified in the Glue application by assigning them to both the relevant departments and company in order to give them accurate permissions. Selected subcontractor staff had also been given access with the edit-restricted permission. While contractor staff were able to see all the issues assigned to each subcontractor, subcontractors were only able to see the issues/documents related to their company.

5.2.2 Model integration process

The native file format that was used to create and update the BIM models was Revit. To publish these models to the iPads, the workflow shown in Figure 5.3 was followed. On-site designers submitted design information through Vault. Off-site designers submitted design information to Buzzsaw. This design information was transferred to the BIM production team through Buzzsaw and they submitted Navisworks models with clash reports. Through workshops, clashes were resolved and the integrated BIM model in Revit was shared with the Glue application directly via Glue Add-in. Because of size and multiple assets in the INA

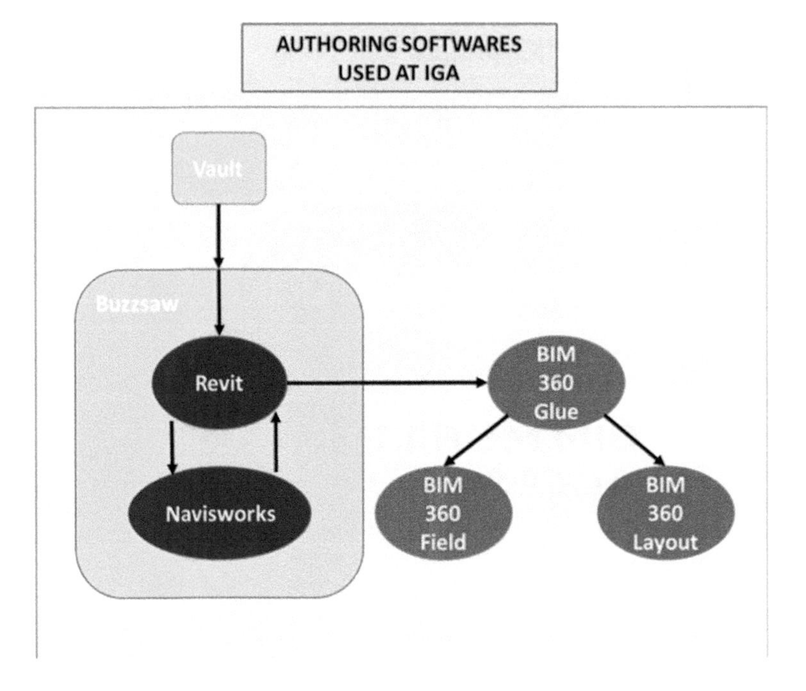

Figure 5.3 Model integration workflow from office to site

project, it was decided to share the BIM models in asset/level/zone-wise logic. Each zone was saved as a view in the Revit models then shared with Glue in the relevant folder created according to the model hierarchy. Discipline- and zone-wise Revit models which were available on BIM 360 Glue, were checked by the BIM production team and by appending all the disciplines in one zone into merged models which were shared with the BIM 360 Field application for site use.

5.2.3 Workflows of digital site engineering

Whilst the design team was progressing with the BIM implementation process and adoption, it was necessary to extend all the efforts performed in the office to the construction site to maximize BIM benefits. With the improved formal workflows enabled by technology and cumulative BIM efforts, the aim was to increase engineering quality, reduce unnecessary rework, and make savings in cost and project schedule.

The mobile BIM workflow ensured that all of the parties were coordinated during the design phase with the clash detection process. Therefore, generated construction documents (shop drawings, BIM model, etc.) were already improved in terms of identifying issues to avoid any possible rework on site. These documents did not just stay in the office, but were also transferred to the site seamlessly via cloud-based mobile applications to the site personnel's tablets to enable BIM-compliant installation management in the construction field. Data flow starting with design information and concluding with the coordinated BIM models and compatible subcontractor drawings is shown in Figure 5.4.

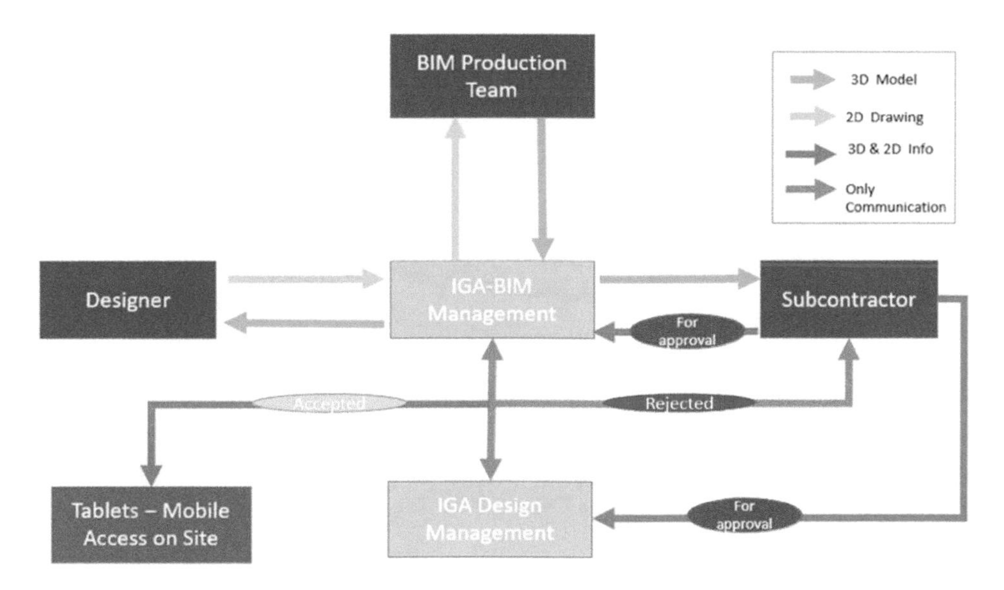

Figure 5.4 Data flow between project participants

Design teams submitted 2D/3D design information to the BIM management team who transferred it to the production team for modelling and integration, then the production team submitted 3D models. Following the coordination process explained in section 3.2.4, the integrated BIM models were circulated to the subcontractors and design teams. The subcontractors produced the construction drawings according to the coordinated models and submitted these to both the BIM management team and the INA design management team for the related discipline. The BIM management team then checked the conformity of the drawings according to the BIM model. If the drawings were approved by both teams, they were shared with the site via tablets, in addition to the coordinated models. If the drawings were rejected, they were sent back to the subcontractor for relevant corrections.

5.2.4 Mobile model access

After finalizing the cross-disciplinary coordination in the BIM environment in the offices, BIM models were shared through the BIM 360 Field application by following the workflow shown in Figure 5.4. Any design change or installation information with integrated disciplines was delivered to site within minutes, as shown in Figure 5.5. This led to faster and correct application of the coordinated design without any delay and with fewer problems on site.

The models were split into zones, which were already created in the system at the beginning of mobile utilization. Two types of updates were identified and carried out to keep the site team informed about the up-to-date BIM model:

Bi-weekly architectural-structural updates:
All the structural and architectural updates were gathered from the related departments in a weekly manner to keep the BIM model up to date. These updates were reflected in the BIM models and shared with the site team bi-weekly.

Figure 5.5 Multidisciplinary merged BIM model view on tablets

Immediate coordination:

Once the cross-disciplinary coordination progress had been finalized, clash-free engineering models were shared with the site team to steer installation. The merged model that was shared with the zone one site (Figure 5.5) included the following:

1. **Structural model:** included all of the concrete and steel elements
2. **Architectural model:** included all of the architectural elements including wall, floor, façade, lifts, elevators, escalators
3. **MEP-IT model:** included all of the coordinated mechanical, electrical and IT services
4. **BHS model:** included all of the elements of the baggage handling system which is the main system in the airport facility.

The resulting clash-free BIM model, in which the necessary engineering decisions considering all disciplines had been made, was shared to the site directly via mobile applications for site engineers to carry out production on site. Observed advantages of design management enabled by mobile BIM applied on site are listed below:

- Any late delivery of design change, which would eventually cause rework and delay on site, was prevented.
- The construction team had the opportunity to foresee their disciplines' further interactions with the rest of the services and proceed accordingly.
- Production on site followed clash-free coordinated BIM models, which led to effective and zero-defect production.
- The site team followed the latest design information with the periodic updates. In this way, physical distance in between the office and site was eliminated.

- Any design and engineering considerations were taken into account while finalizing the BIM models before they were transferred to the site. This led to the correct application of considerations which prevented any further foreseen problems.

5.2.5 Information transfer to the site

Together with the BIM model, 2D drawings compatible with BIM coordination were also installed on tablets after an approval process that included cross-check with the latest BIM model. This control process ensured that everybody was working on the latest and approved design and engineering information. Site engineers did not periodically update the hard-copy 2D drawing that they used to supervise the installation on site. It was very easy for them to use the old revision of a hard-copy drawing since there was no sign that a drawing had expired. On the other hand, this possibility was minimized with BIM-enabled improved information management on site.

Users received immediate notifications after each upload to the project library in the Field application and they could also search through a specific drawing to follow latest revision. These documents could be viewed and notes could also be taken with a mark-up tool. The interface of the BIM 360 Field application site engineers had access to on site via their mobile tablets can be seen in Figure 5.6. Issues, checklists, models, and photos as highlighted in Figure 5.6 are major sources of information that the BIM team and site engineers communicated regularly.

Apart from 2D drawings, other project documents such as material acceptance forms (MAF) and method statements (MS) were also stored using cloud technology, which was accessible through the Field application on site. Without needing any print-outs, all the necessary information needed on site could easily be found using tablets and production was led in the most accurate way. Site engineers could reach the current project documentation on a cloud platform, which made a significant contribution to improved engineering quality. Observed advantages of information management enabled by mobile BIM on site are listed below.

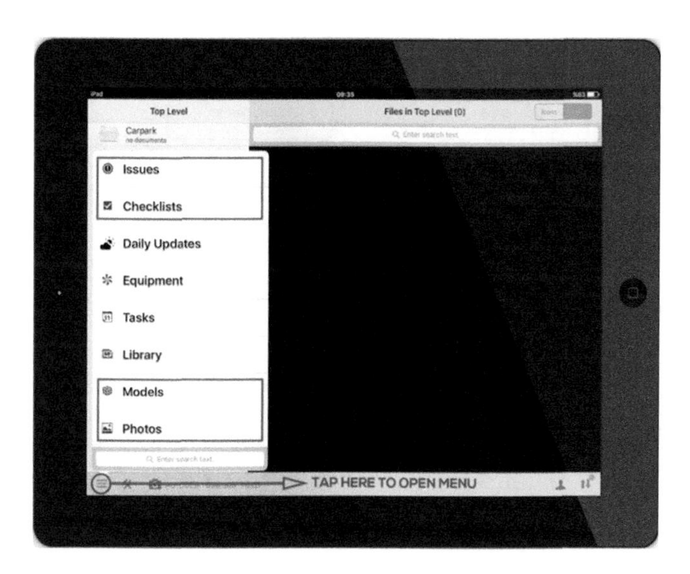

Figure 5.6 Project documents were accessible on tablets

- Any information to be used to manage site works was made available with the mobile devices which could be carried anywhere.
- Production on site followed the most up-to-date and accurate approved project documentation ensuring that desired quality was applied on site.
- Project information did not stay on the papers with old revisions.

5.3 Quality assurance/quality check (QA/QC)

The BIM model was used for any clash detection between the services, allowing problems and conflicts to be resolved before the installation on the site. On-field BIM applications, contrary to traditional QA/QC practice, enabled all QA/QC tasks to be processed digitally on iPads, eliminating the need to wait for parties to sign documents.

A total of 3,210 NFIs were identified in a 1-year period by using BIM 360 Field which brought about tremendous time saving – equivalent to approximately 6,420 man-hours or 802 man-days for the project – which also meant enormous cost savings. This helped the designers and subcontractors to become familiarized with and follow the BIM workflow, which was one of the most important key factors for the project's successful coordination. The BIM MEP-IT coordination workflow can be seen in Figure 5.7.

The mobile BIM workflow shown in Figure 5.7 ensured that all of the parties were coordinated during their design phase. The clash detection process generated construction documents (shop drawings, BIM model, etc.) that were already improved in terms of any issues to avoid any possible rework on site.

While storing QA/QC records digitally, any high-level reporting, which could be generated periodically or spontaneously in various forms to show progress, completeness, trend or details of the defects, was utilized and distributed to the related departments. This drew a clear picture for spotting errors and made it possible to take preventative actions to facilitate zero-error

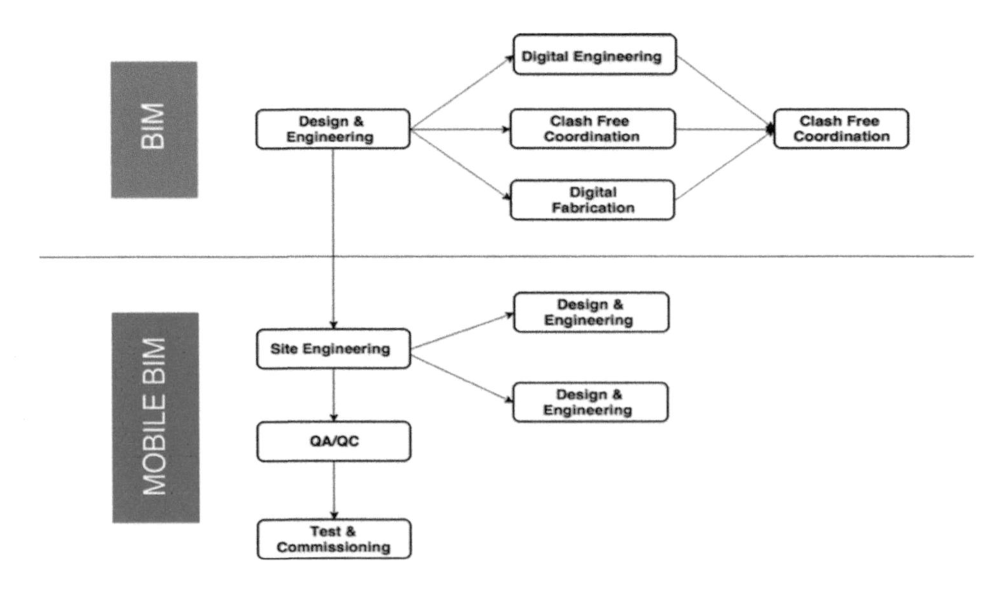

Figure 5.7 Mobile BIM and BIM workflows

production. Since everyday site installation practices were handled and improved with BIM practices, error reduction was made possible.

While setting up the mobile quality system, traditional workflows and checklist documentations were examined and optimized to form an effective process by the INA BIM team. Figure 5.8 shows the improved site inspection workflow enabled by mobile BIM that was put in use by INA to define the NFI process.

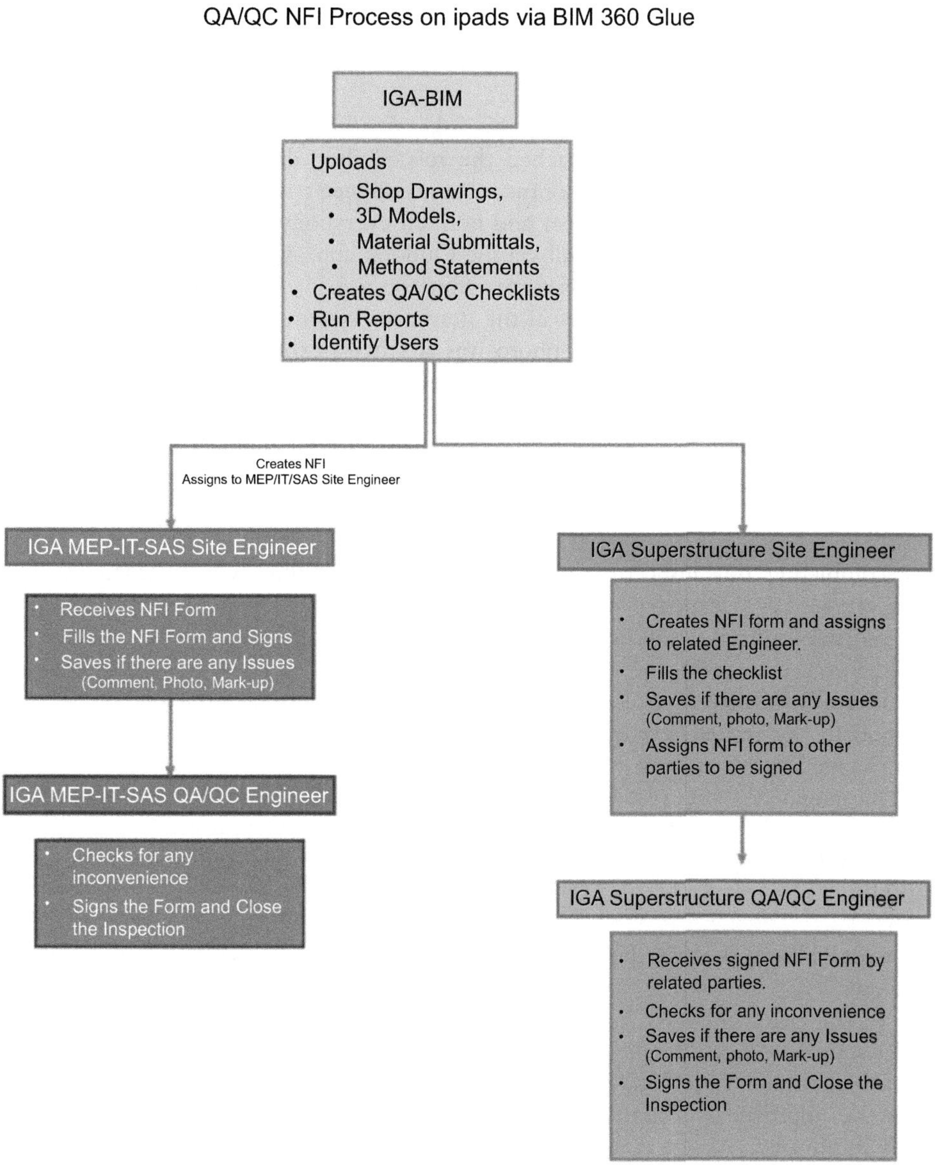

QA/QC NFI Process on ipads via BIM 360 Glue

IGA-BIM

- Uploads
  - Shop Drawings,
  - 3D Models,
  - Material Submittals,
  - Method Statements
- Creates QA/QC Checklists
- Run Reports
- Identify Users

Creates NFI
Assigns to MEP/IT/SAS Site Engineer

IGA MEP-IT-SAS Site Engineer

- Receives NFI Form
- Fills the NFI Form and Signs
- Saves if there are any Issues (Comment, Photo, Mark-up)

IGA MEP-IT-SAS QA/QC Engineer

- Checks for any inconvenience
- Signs the Form and Close the Inspection

IGA Superstructure Site Engineer

- Creates NFI form and assigns to related Engineer.
- Fills the checklist
- Saves if there are any Issues (Comment, photo, Mark-up)
- Assigns NFI form to other parties to be signed

IGA Superstructure QA/QC Engineer

- Receives signed NFI Form by related parties.
- Checks for any inconvenience
- Saves if there are any Issues (Comment, photo, Mark-up)
- Signs the Form and Close the Inspection

Figure 5.8 Improved QA/QC workflow enabled by mobile BIM

According to the improved workflow, the first step was to identify all of the QA/QC checklists to be used during inspections in the BIM 360 Field application, either manually or using Excel spreadsheets. After creation of the checklists on the system, a template was also saved for future assignments for each task type for easy reference by tablet users. The steps of the inspection workflow were standardized in detail and users were all trained in the same manner to avoid variations during the practical implementation.

This standard procedure that was created and put in use is explained in detail below:

- The subcontractor notified the contractor on a daily basis when the site installation was completed for each zone.
- The contractor QA/QC team scheduled an inspection programme according to subcontractor notification and sent the information (zone, discipline, date and time) to the BIM management team.
- The BIM management team created daily tasks according to the schedule provided by the QA/QC team, to which they attached the related shop drawings and the checklist form which was filled in with the subcontractor and assignee's details, and finally they assigned the tasks to the relevant site engineer and notified everybody via email.
- The site engineer received the email via the Outlook app, synchronized his tablet to receive his tasks and filled in the inspection form offline on site.
- If there were any non-conformities in the items, the system created an "issue" and assigned it to the relevant subcontractor. A photo was attached (if the site engineers had taken any), or a document that was already uploaded to the Field application or, alternatively, the BIM model was marked-up.
- The site engineer assigned the completed task to a QA/QC engineer.
- The QA/QC engineer checked the inspection form and gave an overall rating and either closed it or left it open.

After completing this inspection workflow, since everything was recorded in the cloud system spontaneously, corrective actions were taken immediately. The corrective action procedure was also standardized as follows:

- Open items identified by site engineers were shown in the subcontractors' web-based dashboard with the related attachment. After the subcontractor completed the rejected work, they changed the item status to "work completed".
- The site engineer who created the issue was automatically informed, and confirmed if the work was completed correctly and changed the item status to approved.
- The QA/QC engineer was automatically informed, and confirmed and closed the checklist.

Advantages observed with the implementation of QA/QC enabled by mobile BIM are listed below:

- Collaboration of both people and parties has been achieved in digital platforms. This has eased the communication and provided quick and effective resolutions on site.
- With the improved workflow of inspections, time spend on this process has been significantly minimized.
- This system implementation made it possible to transfer all of the data stored during construction to the next processes.

Figure 5.9 Use of BIM 360 on site

- No time has been spent on paperwork which has avoided any negative effect on project delivery.
- Any non-conformance trends have been detected and necessary actions have been taken to address these issues all together before they could have any major impact in terms of delay, rework and quality.

The 3D/4D/5D modelling and coordination environment was provided by the BIM authoring and the 4D and 5D simulation tools. The most outstanding advantage of the coordination workflow was the real-time progress control, monitoring and timely communication of the quality checks and control on site via the 150 iPads, including all the coordinated BIM models used by the site engineers.

BIM 360 Layout and compatible robotic total stations were utilized in the project by the BIM management team to closely track BIM conformity of the site installations. Total stations fully compatible with Autodesk BIM 360 layout were procured and used precisely. Figure 5.9 shows an example of this real-time project progress control and quality checking.

5.4 Observations

An observation issue was registered when the BIM engineers on site found a problem in a constructed area. Before the inspection could move on, the observation issues had to be resolved by the responsible company. This enabled minor inconsistencies to be addressed before they could cause further major problems.

If the site BIM engineers discovered any problem with on-going on site construction work, they created a supervision issue from their tablet in order to address the problem before the work was completed. It enabled the problem to be resolved one step before observation, without causing any rework. If the problem could be resolved before the work was completed, it was converted to an observation issue.

5.4.1 Site inspections

When the construction was completed with no supervision or observation issues, inspection forms were registered with tablets by INA site engineers. Any issues identified during inspection in terms of QA/QC were recorded digitally and assigned directly to the related party to be resolved. Completion percentage commissioning was directly updated in the BIM system which could then be followed.

5.5 Progress monitoring

Issues regarding the construction progress according to the baseline were detected and registered by the BIM engineers on site as progress monitoring. Since they had the completed view of the construction through the BIM model, they were able to determine the work sequence and find the problems.

An enormous 4D model was developed that integrated more than 30,000 activities to enable daily and monthly monitoring of the INA construction progress, giving the construction teams dynamic control over progress in the INA project. For the purpose of executive decision-making, 4D snapshots of the key assets and zones were shared weekly with directors and the responsible bodies. In addition, the 4D BIM model was used for coordination and communication with the related stakeholders who also attended the BIM room meetings.

The main advantage of BIM use for quantities management was the ability to derive actual quantities of items used in the project to eliminate any excess amount and verify the estimated amount. Bills of quantities (BoQs) from the BIM model were generated and issued to the technical offices of each discipline. Figure 5.10 shows the workflow of 4D/5D integration with the site progress reporting.

The workflow shown in Figure 5.10 explains how BIM-based project progress monitoring applied in INA. Since planning, quantities and QA/QC were directly carried out with BIM and the progress payments were done accordingly, subcontractors had to comply with BIM both for their work on site and also reporting to INA. This meant they had to follow the milestones given in the baseline, install on site by conforming to QA/QC standards together with the BIM model, and report the quantities in line with the BoQs extracted from the BIM model.

Since the airport is to be operated for 25 years after its completion by the INA consortium, BIM will play a crucial role not only through the design and construction stages but also during the operation stage. In other words, the procurement philosophy and project requirements make BIM use critically important for cost effective and sustainable facilities management of the airport infrastructure.

5.5.1 Site survey with BIM

Autodesk Point Layout was used in the INA project. It is a point creation plugin and is part of the Autodesk BIM 360 Glue project-based offering that integrates Autodesk Navisworks, AutoCAD, and Revit software to help make construction field layout, QA/QC and as-built modelling faster and more accurate by direct export/import of BIM data to and from the field.

The BIM site engineers were assigned to not only track the resources on site but also monitor the performance of the project. Site engineers recorded progress issues by attaching the

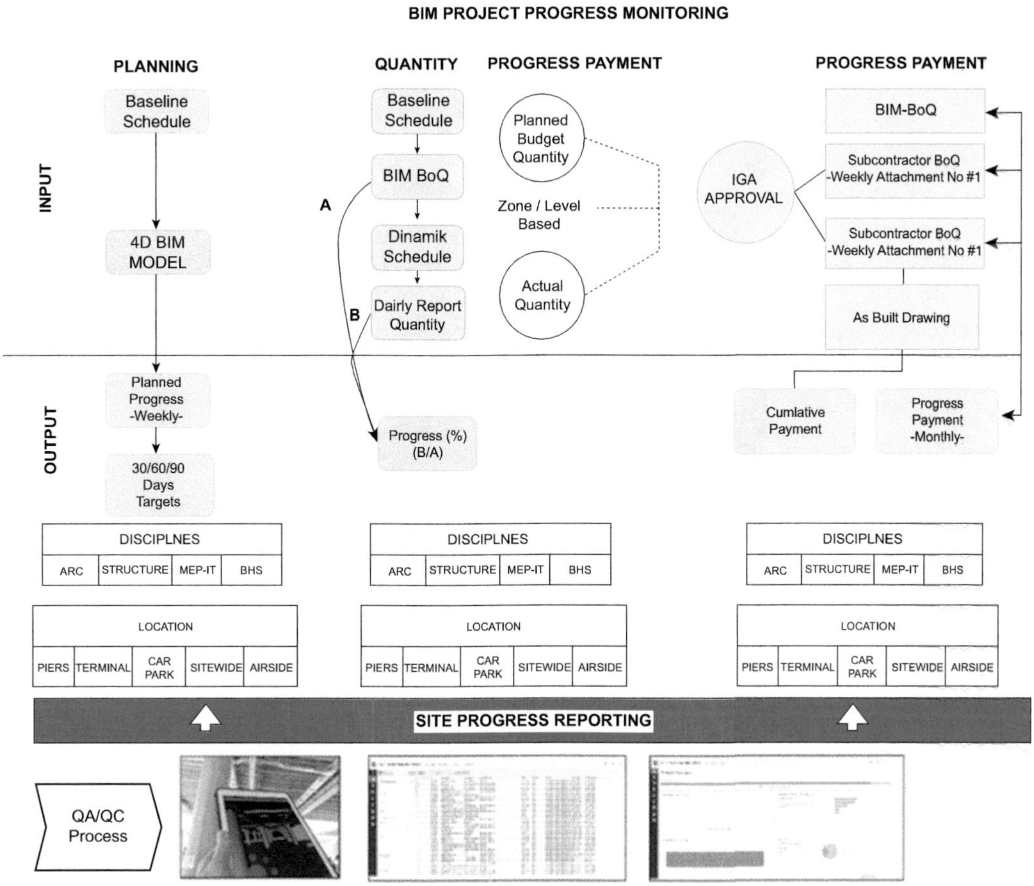

Figure 5.10 BIM workflow for progress monitoring

relevant information from their tablets with the BIM 360 Field application. Both detailed reports for discussion in the weekly meetings and summary reports for presentation at the executive level were generated to monitor progress and related issues. The most important outcome of this monitor reporting was observed as having access to real-time information which was automatically updated directly from the construction site. Subcontractors were allowed to see their progress in terms of the completed inspections, observations and also progress issues.

The BIM site team was able to record any as-built information digitally when there was any variation from the BIM model. Since these records stored asset, zone, site photos and related company information, they were accessible to anyone in the project and ready to report with the required filters. To control critical site installations precisely, the BIM 360 Layout application, which was compatible with the robotic total stations and BIM models, was used by the BIM site team as shown in Figure 5.11.

Site control points required by the BIM management team were created in Revit models and shared with the Glue application for measurement via the Layout application to identify

Figure 5.11 Controlling site installation with BIM 360 Layout application

any deviations. Advantages observed with the performance management implementation on site are listed below:

- The executive level was informed using up-to-date reports generated through the BIM system to avoid spending time on the unstructured, detailed and out-of-date reports. Management decisions were taken by analysing BIM-originated reports.
- Spontaneous reporting was possible by including any required information.
- No effort was required to gather the various forms of data that the different departments were holding to create reports for monitoring the project.
- The contractor had immediate access to the subcontractors' performance reports, which provided a way of effectively managing them to deliver the project in a timely manner.
- Subcontractors were able to monitor their progress in terms of completed tasks, not-approved works and approved shop drawings. In this way, they were able to take corrective actions based directly on this system.
- Various subcontractors performing the work on site had access to the latest project information to be followed.

5.6 Conclusion

Enabling mobile BIM was one of the core competencies of the INA BIM team. The processes defined for mobile BIM evolved throughout the construction period as new digital tools and plug-ins were integrated with the core BIM 360 Field and BIM 360 Glue cloud applications. Project control and monitoring was managed via the mobile BIM implementations, and the rate of aligning the learning curve of site engineers with the BIM team was significantly high.

User-friendly interfaces of the mobile BIM applications, well-defined information flows, easy access to project data, and easy communication led to the high adoption rates on site. To conclude, mobile BIM is a must to conduct and manage digital site engineering, observation, QA/QC, and progress monitoring for such a mega project.

6 Key learnings about ABIM and paving the way for the airport operations

CONTENTS

6.1 Introduction

The architecture, engineering and construction (AEC) sector has been facing considerable challenges recently due to the increasing scale and complexity of projects. Mega projects are more difficult to manage in terms of decreasing cost and increasing quality and productivity. Innovative approaches have been proposed to overcome the various challenges faced by the AEC sector. Achieving integration, and thereby a more collaborative project environment, is essential in this process. Today's key trend in successful business strategy is put as "combine and conquer", which includes innovating business models together with transforming the core engineering systems around digital technology.

Accordingly, a rapid increase in the implementation of Building Information Modelling (BIM) in mega projects has been observed. BIM provides significant increase in efficiency of project execution through optimizing the project constraints of scope, time, cost, quality and resources. Therefore, incentivizing all project parties to work in a collaborative fashion can be considered an important key success factor.

Overall, Chapter 6 will convey the experience and learnings of BIM use in airport construction gained through the INA project, and how the INA BIM team used this improved understanding to pave the way for the airport operations. Holistically, the process of Airport BIM (ABIM) implementations can be interpreted as innovation diffusion. The rate of this innovation diffusion was influenced by the enablers and challenges that the INA BIM team faced throughout the construction phase of the project.

6.2 Enablers and challenges of the ABIM implementation

There were mainly two categories of problems that were encountered in the INA project, including engineering and managerial issues (see Figure 6.1). These two categories of challenge

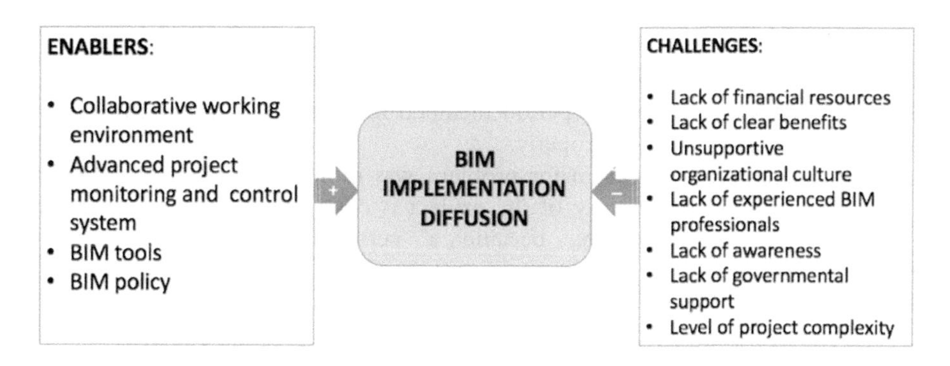

Figure 6.1 Realized enablers and challenges for BIM implementation diffusion

included a lack of experienced BIM professionals, the level of project complexity, and lack of awareness.

Regarding engineering problems, clash resolution processes related to coordination between mechanical, electrical and plumbing (MEP) systems and special airport systems (SAS) have become one of the major issues, in both the design and construction phases concerning a wide variety of project individuals. Managing the flow of requests for information (RFIs) and incorporating the solutions, which have been generated from different discipline perspectives, has represented a major engineering management problem.

An airport project, due to its nature, requires different and much more complex types of mechanical systems that need to be placed in large areas and activated altogether. Figure 6.2 indicates the challenge of the INA project's complexity, as there is a cluster of various types of pipe, duct and cable tray systems (e.g. HVAC ducting, plumbing pipes, fire sprinklers, electrical and IT cable trays, and heating and cooling pipes) at different levels of the terminal building which required accurate coordination to aid practicality on site.

Moreover, BHS placement has been a significant engineering challenge in the INA project due to the requisite accuracy and the length (42 km) of the baggage routing. Initial engineering decisions were made based upon the BHS systems' placement. Then, MEP systems including

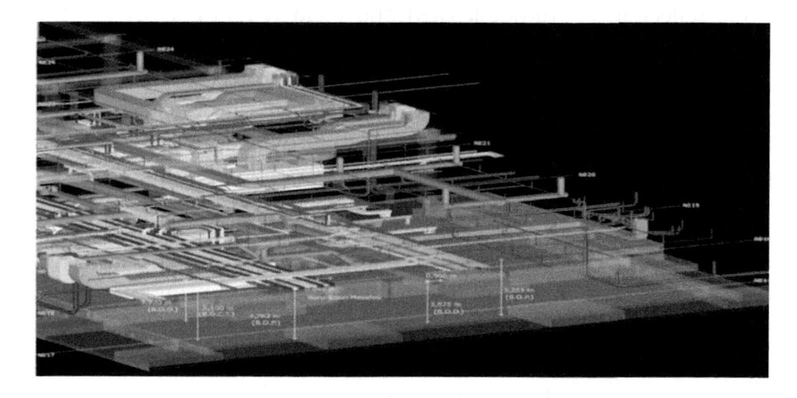

Figure 6.2 Viewpoint of MEP systems

HVAC ducting, piping, electrical and IT cable trays were placed appropriately in an architectural and structural envelope, and coordinated accordingly. However, because the MEP subcontractors had limited experience in making interdependent disciplinary decisions in such a large-scale project, the coordination period included many conflicting iterative processes that needed to be defined and managed properly.

Regarding managerial issues, the major problem was monitoring and controlling work on site. As far as the size and complexity of the project is concerned, managing all project individuals, mainly the subcontractors, became a very challenging issue that required a substantial management plan. In the very beginning of the INA project, the subcontractors' lack of awareness and experience, and resulting resilient attitude against engaging BIM processes in their daily site and office work, created a necessity to train all subcontractors through facilitated workshops.

To overcome the challenges faced throughout the implementation of BIM in this mega airport project, strategic control mechanisms including periodic BIM workshops for educational purposes were employed. It is essential here to demonstrate how BIM was taken over to the subcontractor on site and how BIM led to the installation work of the subcontractor on site. Control mechanisms were provided by the predetermined BIM execution plan and strategy, and also workflows for each of the BIM processes at the very beginning of the project; and via these mechanisms, subcontractors were fully integrated into the BIM environment.

The BIM department that was represented as the BIM management team was responsible for managing, integrating, and monitoring and controlling the BIM model data input from the project subcontractors of various airport design disciplines. BIM models were generated with different levels of detail and the concurrent engineering and design proceeded. Clash reports, 4D scheduling and QA/QC reporting on site were provided by the BIM department to provide an effective control mechanism on the subcontractors' work. Weekly BIM workshops, and BIM coordination meetings were used as communication tools to oblige subcontractors to use BIM tools. The use of this BIM software enabled the INA project individuals to have controlled work sharing, BIM coordination, design review, change visualization, quality and issue management, access to RFIs and submittals, and notification of inspection documents.

The BIM policy of the company declared strict contractual obligations for all subcontractors to make them follow and utilize the BIM process in their work processes, such as using mobile tablets on site for filling out notification for inspection (NFI) documents to get their progress payment. All coordination issues were detected on site by the client's BIM site engineers for each manufactured zone.

The issues were reflected on Autodesk BIM 360 Field system periodically to track each subcontractors' performance on site. These reports were shared internally weekly so that BIM processes enhanced the control mechanism. Accordingly, project parties that consisted of the designers and subcontractors were familiarized with using the products of BIM in a harmonized fashion. For instance, on the site, 150 mobile tablets provided all coordinated BIM models to the site engineers to assist them in carrying out zone-wise production. Apart from 3D models, approved 2D shop drawings were also provided for the field via mobile tablets.

Along with the production on site, QA/QC was conducted with the help of digital documentation which supported cloud-based access for all related site engineers. All of these applications took place on the Autodesk 360 Field platform. Additionally, a 4D model including 30,000 activities was generated to track progress on a daily and monthly basis to enable dynamic control over the progress of the project.

Disciplined and zone-wise clash detection was utilized throughout the design and construction phases. The frequency of clash detection and resolutions depended upon the frequency of design revisions. The airport systems integration was dynamically controlled via periodic clash detection. The periodicity was determined by the submission schedule of subcontractors. However, the BIM department determined and controlled the coordination process of mechanical, electrical, plumbing and information technologies (MEP-IT) systems with a separate coordination workflow due to their highly complex nature in such a mega-scale airport project.

The workflow depicted concurrent engineering and design in a fast-track fashion and the responsible parties in the process. The main objective was to resolve the clashes at the LOD 350 BIM level with MEP designers and proceed to the extraction of shop drawings out of the clash-free BIM model to push the work on site. The BIM model was continuously updated with various details such as equipment details and specifications throughout the workflow. Every update to the BIM models and shop drawings was shared in the cloud system and made accessible via mobile tablets on site.

Implementation of BIM in airport projects is significantly different from the typical applications of BIM to new building construction in which the focus is on the design and construction of a lone building. BIM use for airport construction requires more complex BIM implementations compared to buildings, because the airport design and construction incorporate a varying mix of infrastructures including terminals, runways, passenger gates, car parks and transportation systems including railways and roads. An airport construction project therefore comprehensively covers all aspects of these different construction types.

In the INA case study, BIM was used from the early briefing and concept design through the detailed design and construction phases. Since the project stakeholders recognized the crucial benefits and advantages of using BIM that make concurrent design and construction possible, it was decided to also use BIM in the facility management phase after the completion of the construction. Essentially, it was realized that BIM had a significant impact on the following matters: it provided authority over subcontractors while managing the work and delivering on site; improved the quality of the design and construction stages; reduced waste both on site and in the office; enabled fast resolution of issues on site; and enhanced collaborative work.

6.3 Motive to innovate

It can be stated that one of the major challenges of the INA project was completing the project within a very limited timeframe. The common idea was that the activities should be completely correctly without needing any rework. Processes that did not bring any value to the project were to be minimized by using technological tools and devices. These processes included paperwork for registering any type of data, approval efforts and such like.

Some of project end-users mentioned about the challenge of working on such a large construction and also the long distance between the site and office. Collecting project documents such as construction drawings or QA/QC documentation on a daily basis would have been ineffective. It would have also created delays or mistakes in reaching the most up-to-date and accurate documentation throughout the project because design and construction were progressing in parallel. Easy, fast and effective ways of accessing the documents were highly needed.

Because of the complexity of the project, it was predicted that there would be many congested areas where different services built by various departments would interact with each other. Efficient and effective communication between the project participant companies and departments was achieved with the common environment that BIM provided.

6.4 Innovation process

Questions regarding the innovation process were directed only to Site BIM engineers and the BIM director because they were responsible for managing or conducting the innovation. First of all, along with the BIM engineers in the office, a site BIM engineer team was set up in order to carry out the digital transformation from office to site. In the beginning, five engineers were hired and assigned to the different locations of the construction area. They were responsible for implementing BIM functions on site which also included adoption of site engineers to the new system.

The most important element in the process was stated as being the daily follow-up of activities on site by BIM site personnel. They made personal communication with site engineers and provided effective communication with them. It can be pointed out that at the very beginning of the project, only a few of the site personnel were eager to use BIM on site. That is why BIM site engineers had to identify the key contacts in their area who had more interest and make contactwith them first. The common initial idea of the site personnel was that they would continue to perform daily construction activities in the traditional way and use BIM as a control mechanism in case of a problem. It was the first BIM experience for the site engineers and they were therefore not aware of the capabilities of BIM. Site BIM engineers had to change this perspective and make sure tablets were actively used during installation.

The innovation was not fully accepted in the first month of implementation and site teams did not actively use their tablets for installation or for digital NFI registration. The BIM site team had put pressure on the subcontractors and also on the INA site team, as will be explained in the following sections. Site BIM engineers tracked the works on site on a daily basis and gave notice to any personnel who were not using their mobile devices. It was also noted by some end-users that BIM implementation would not be successful if there was no effective strict control mechanism on subcontractors.

6.5 Adoption strategies for the use of BIM on site

During the adoption phase of the BIM system by the INA site engineers, their main argument was that they knew how to do their job in the best way and that the site BIM engineers did not have enough experience to change this. They had an emotional reaction when someone from outside their department (site BIM engineers) intervened with their construction works.

Site BIM engineers were responsible for checking the work on site and comparing it with the coordinated BIM model which was accessible on site via mobile tablets. However, during such quality checking, many debates occurred between the BIM site team and subcontractors on site. Accordingly, coordinators of the relevant departments were instructed to follow BIM very strictly. The significance of BIM was even continuously secured and expressed by the CEO of the INA project.

When two or more services physically interacted on site, problems occurred between parties which they failed to resolve very quickly. There were only verbal arguments followed by old fashioned, inconsistent and uncoordinated construction documentation. Therefore, they had to refer to the BIM coordination model for accurate and quick solutions. Because the BIM model constituted the most updated revision and the coordinated services, it was very straightforward for parties to identify what had caused the problem on site and which action should be taken and by whom in order to correct it. Once they had experienced this opportunity, they started to figure out the advantages of using BIM during construction.

Use of BIM was diffused from top to bottom. Before the start of construction, site personnel agreed what should be in the BIM model to avoid rework. Site personnel of INA were grateful for BIM as it meant construction only had to be performed once and they were sure of how the end product would come out.

Both personal pressure from the BIM site engineers to the site staff who were performing the physical work on site and also managerial pressure from the BIM coordinator to other department coordinators played an important role in the adoption process. The site personnel experienced the benefits of using BIM themselves and this led to a transformation from the traditional method to the innovative ways of BIM.

Enforcement of the use of BIM for resistant subcontractors was managed in two ways. The first one was that in order to claim the monthly payment from INA, subcontractors had to document digital inspections forms as approved. If the inspection forms were done on paper and not digitally through mobile devices on site, the payment was not approved by INA. This obliged subcontractors to perform inspections digitally in order to receive the payments entirely.

The second way was quality control of the constructed area according to coordinated BIM models. Site BIM engineers created observations from their tablets if there is any incompliance in the constructed area. When the subcontractor requested inspection of the manufactured zone, the inspection form was not even registered because of the on-going observation issue. Subcontractors had to resolve the observation issue first and then request the inspection later. Since their monthly payments were based on approved inspections, they had to follow BIM coordination and resolve any incompliances.

Despite these two strict procedures, some of the subcontractors failed to follow them in some areas. Once they faced the fact that they would have to comply with BIM in order to receive the payments, they had to do some rework which cost them money and time. With these experiences they had to accept the BIM system and do their work accordingly.

6.5.1 Main advantages and disadvantages of the use of BIM on site

The INA project team had access to the most up-to-date and coordinated project information. With only 2D construction drawings it would have been very challenging to construct such a complex project. Many times, construction drawings were too complicated to understand and transfer to the installation. Conversely, performing construction was quite straightforward with coordinated BIM models. The main advantages of using BIM models over drawings were listed as being able to (1) take measurement from any reference point, (2) see the adjacent systems coordinated, (3) access the latest revision and (4) determine the method of construction. These functions enabled the INA project team to make progress with construction without reworks. Since the consolidated BIM model was available before starting the construction in an area, the best method and sequence for carrying out the construction had been decided by the site team and workers were instructed accordingly.

The BIM system provided a common digital environment for the parties to address and resolve the issues in a very transparent way. Fast access to project documentation when needed was provided with the use of the BIM system. Time-consuming formal procedures for construction were eliminated through supervision of BIM.

Subcontractors pointed out that the ability to access digital documentation any time eased their track on the project in terms of their performance. They did not spend time on repetitive paperwork. The disadvantage for the site engineers was the necessity to learn a new technology and get used to it in a short time. In addition, site BIM engineers dominantly intervened in the site engineers' work. In the end, these efforts brought a lot of value to the end product. Without the BIM system it would not have been possible and they would have faced very serious issues.

A disadvantage from the perspective of the site BIM engineers was that they had to spend a lot of time training and helping site staff adapt to the new system. They had to remedy the unwelcoming response to their existence because of the site personnel's resistance to change.

6.5.2 Grounds for innovation adoption

The most important reason for successful adoption was to have the on site BIM team work in separate areas in the project. This team was involved in the daily construction activities and followed the work on their area very closely to address any issues before they could occur. The successful communication of the on site BIM team affected the adoption in a very positive way. They were able to be part of the site team with their efforts and their contributions.

Leadership commitment on BIM had a big impact on the project team's willingness to adopt the new system. The system was well supported by the CEO of the main contractor. The numbers of tablet was large enough for them to be used on site effectively by all of the departments in the different construction zones.

6.6 What is next: paving the way for the airport operations

Smart airports utilize new technologies to improve end-user experiences while ensuring economic feasibility and aeronautical safety. Building Information Modelling allows managing and accessing big data encompassing physical and operational data of an airport. Creating a comprehensive BIM implementation framework that leverages use of big data enables digital transformation of tasks involved in the management of airport projects throughout their lifecycles. Such a transformation allows assets to be built, maintained, and operated in a smarter way.

According to a market report on the business value of BIM for infrastructure (Jones and Bernstein, 2012), while 50% of infrastructure projects are using BIM for 1–2 years, less than a quarter of them (23%) feature BIM use for more than 5 years. Similar results are observed in airport projects. Since BIM is mostly adopted for the design and construction phases, such projects do not result in a complete digital transformation from a whole lifecycle perspective. Consequently, one of the objectives of the INA ABIM implementations throughout the airport lifecycle was to enable smart airport facility management for all stakeholders.

Through the airport operations, the end-user and passengers become the key stakeholders. Airports are, after all, business ventures and so must be able to make money from the airlines that use them and the passengers that travel through. Facility management is the phase in which competency in digital disruption becomes of utmost importance. Via successful facility management, INA aims to handover and then manage the airport's data via BIM without any loss, contradiction and misinterpretation of data.

6.6.1 Strategy for BIM for the airport operations

The INA strategy is based upon utilizing an Airport BIM (ABIM) management framework to provide a generic strategy for BIM data handover from the design and construction phase to the operation phase (6D BIM). Smart airports are just a part of the smart city concept. There are only a few companies who work on the smooth integration of systems to create an end-to-end journey.

However, in such a fragmented sector like aviation, integration of systems and management approaches is still a great challenge, and not yet solved.

The Istanbul New Airport project integrated with 6D BIM for facility management. There were four steps for the BIM 6D:

1. Creation of commissioning data and development
2. Development of a commissioning and operational model
3. Airport system and integrated model information
4. Completion and handover

At the first step, creation of commissioning data, INA checked and controlled over 12,000 approved SD and IFCs which were already shared in the common data environment BIM 360 Field.

As part of the first step, a room tagging strategy was developed, and room lists for all of the buildings in the project were collected. In order to create the tagging, sequence number, room names, and disciplines were arranged. With the combination of this information, room tagging for each room was specified and ready to be updated in the model. The room tagging strategy is depicted in Figure 6.3.

As can be seen in Figure 6.3, a combination of the discipline of the relevant system, room name in English, and sequence number were used respectively to come up with the appropriate room tag.

BIM engineers classified and verified over 8,000 approved shop drawings considering all of the design systems and assets in order to develop the BIM 6D system (Figure 6.4). Classified and verified shop drawings were used for 3D commissioning and operational modelling.

In addition, the creation of equipment family categories and types was processed via examination of the shop drawings, equipment lists and BIM BOQ tables. The equipment was categorized according to its discipline and operating system. Hence, categorization of all operating systems available in the BIM model was essential (see Figure 6.5). Five main systems were filtered: mechanical, electrical, extra low voltage (ELV), information technologies (IT), and special airport systems (SAS).

Each system had its own subsystems which were also defined in the BIM model. Equipment that was associated with these operating subsystems was also categorized. The equipment family

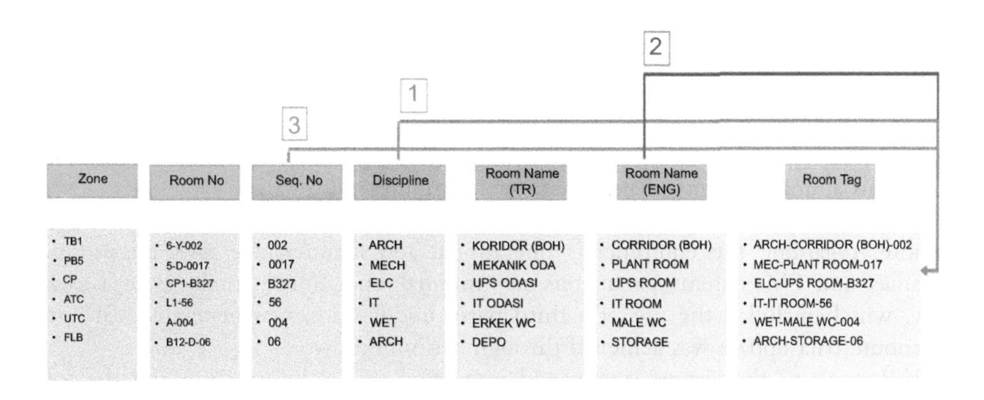

Figure 6.3 Room tagging strategy

System Type	Number of SD
02-Low Voltage System	139
04-Lighting System	216
05-Lighting Automation System	204
06-Fire Detection	160
07-Public Alarm & Voice Address System	475
08-Small Power System	182
10-Cable Tray Plans	3
14-BMS System	640
15-Energy Metering System	73
17-Utility Plans	35
18-Emergency Exit Lighting	325
Panel Installation Plans	75
Passenger Toilet Socket Layout	411
WC Typical System Projects	22
ULV Room Details	1
3.Grounding&Lighting Automation System	2
Mechanical Room Details	147
TOTAL	**3110**

Figure 6.4 An example table showing the number of shop drawings (SD) for each different system

category list regulates data acquisition for the 6D model application in operations control and maintenance during the facilities management practices.

Materials approval forms were also classified via tracking on a daily basis to create commissioning data. The second step was the development of a commissioning model. The commissioning data collected during the first phase was progressively incorporated into the BIM model to develop the operational model. This information included the following items:

- Room tagging
- Mechanical equipment details assignment (tagging, dimension, family system assignment)
- Main and second fix electrical equipment modelling from the approved shop drawings
- IT equipment incorporation as a 3D model or from the approved shop drawings
- LET equipment tagging
- Architectural suspended ceiling plans
- Operations and maintenance data (equipment attributes)

Furthermore, regarding the equipment's operational and maintenance data, to produce lean family names and FM systems parameters, an internal operational family system assignment workflow, which included the use of a third party middleware, was generated. Bi-directional model attribute data update was achieved through this workflow.

The third step was the airport system and integrated model information and the following actions were considered:

SYSTEMS				
MECHANICAL	ELECTRICAL	ELV	IT	SAS
Heating System	High Voltage Systems	Fire Alarm and Detection System	Access Control System	Baggage Handling System
Potable Water	Low Voltage Systems	Energy Measurement System	Passenger Baggage Drop System	Passenger Boarding Bridge
Waste Water Systems	Medium Voltage Systems	Lighting Automation System	Car Park Guidance System	Ground Supporting Equipment (Ground Power Unit)
Fire Fighting System	Uninterruptible Power Supply Systems	Pier Gas & Leak Detection System	Car Park Management System	Pre-cooled Air
Domestic Water	Automatics Transfer System	Advanced Surface Movement Guidance and Control System	Common Use Self-Service System	Visual Docking Guidance System
Cooling System	Lighting System	Public Address and Voice Alarm System	Flight Information Display System Electronic Monitor	Explosive Detection System
HVAC	Airfield Ground Lighting System	Building Automation System	Master Clock System	Security Screening System
Plumbing System	Load Shedding System		Perimeter Security System	Automated People Mover
Fuel Hydrant System			Video Surveillance System	Lift , Escalator, Travelator (LET)
Waste Collection System			MATV System	
Sea Water Pumping System			Passenger Security Screening System	
Sanitary Systems – Drainage Systems			Information Communication System	
Irrigation System			WiFi Systems	
Fuel Automation System				
Aviation Fuel System				

Figure 6.5 Operating systems categorization

- Navigation of the BIM model and viewing of items of interest (equipment, meeting rooms, incidents etc.), along with information (by way of cards/widgets)
- Performing monitoring and control of equipment
- Incident reporting via mobile applications and showing the same on smart BIM
- Meeting room reservation via smart BIM
- Searching for items of interest (equipment, location, incidents, meeting rooms etc.) and viewing on smart BIM

The final step was handover, which has now been accomplished.

6.7 Conclusion

Transforming the INA project around digital technology was achieved by integrating key technology and people in one virtual BIM platform. Embracing the digital transformation strategy at an early stage enabled collaboration management, and execution of the project on time and budget. However, from the outset, it was observed that the learning curve was very steep for many of the stakeholders which challenged the integrated project delivery system. A significant amount of challenges were overcome early on through mobile BIM use, using BIM tools at their full capabilities, and efficient and fast information sharing.

In a more detailed manner, there are some technical realizations to sum up. It was realized that in the congested areas the BIM model could be very complicated and that a progress tracking function should be added to the mobile applications. Furthermore, the significance of BIM should be determined by an organization schema of the company to direct other departments from the beginning of the project. Authority to enforce other departments should be given to the BIM department by the organization schema.

Overall, the INA project staff stated that they entirely understood the importance of using BIM after experiencing it, and success in fast adoption of BIM in such a mega project led to the idea to develop strategies for BIM use in the airport operations and maintenance phase.

Reference

Jones, A.S. and Bernstein, H.M. (2012) *The Business Value of BIM for Infrastructure*, https://images.autodesk.com/adsk/files/business_value_of_bim_for_infrastructure_smartmarket_report__2012.pdf.

7 Conclusion

CONTENTS

This chapter summarizes how lifecycle ABIM was initiated and rolled out by unveiling key insights at a strategic level and touching upon a more executive perspective of ABIM implementation processes.

7.1 Key features for creating a competitive edge for airport operations

Airports are very complex infrastructures, far more complex than any other infrastructure construction projects. It is a massive construction type utilizing large-scale technologies and integrating complex ecosystems. The best way of managing the airport operations is to understand how the whole place and the system work together. This task has become more challenging as airports' key design and construction features have changed drastically in recent decades. Design, engineering and construction have been fully integrated, and procurement methodologies have been significantly renewed. Additionally, new technologies should be followed closely, and applied for commercialization purposes.

Implementing smart airport systems can also bring significant competitiveness to airports. However, because of uncertainty about what and how to deploy such sophisticated technologies, airports may suffer a substantial loss in operations which pushes them back. How can they scale their services in accordance with the changing number of passengers? It is a very complicated system in which there is no single formula, and operator teams might fall short in terms of understanding how complicated that system is. As they cannot align their investments, they are afraid of investing in such smart/new systems.

7.2 Key ABIM strategies differentiating the INA project

Bringing all stakeholders (passengers, airlines, operators etc.) together on the same page in a peaceful way will bring a competitive edge to an airport. Thiscompetitiveness is not about how spacious and mega the structure is. Without utilizing an integrated way of delivery – procurement strategies to align stakeholders' interests – it could take a minimum 5–10-year timeframe to make an average airport project operational.

In the case of the INA project, utilizing a strategy to follow an end-to-end fully digitized approach with a client representative mindset brought a competitive edge. Airport Building Information Modelling (ABIM) implementation is the approach that enabled delivery of the whole project lifecycle on behalf of the client. That being said, the integrated project delivery (IPD) mindset that led to a fully seamless delivery with a client-representative role was achieved through the BIM digital platform. Beyond execution, within the start-to-end BIM delivery, which was a huge journey of seamless execution, requirements were very well defined and internalized by the BIM team on behalf of the client. All project teams delivered the project as one team by utilizing a seamless digital platform. The key success driver was strategizing how to translate this complicated engineering process to construction.

Different airport projects across the world can give insights on different project phases that are managed and delivered through a BIM ecosystem, but INA is unique in terms of managing and delivering all lifecycle phases by a seamless BIM implementation. The mindset was created by asking "What can the client achieve beyond this?", "What is important to the client?" It is known that the client needs zero defects on site with no cost over-run without losing time and quality.

7.3 Customizing ABIM implementation principles for the INA project

Customizing the strategy by the motivation to satisfy the client's requirements is also crucial. All parties should share this motivation and digital is the only solution for all the parties. Accordingly, the scoping phase of the INA project took 3–4 months. Later, the platform was launched. Overall, there were lots of mechanisms working behind the ABIM platform and it took 3–4 months to integrate and develop these. Furthermore, due to the nature of the INA project, which required dealing with human behaviour change, the predetermined mechanisms were updated and adopted accordingly over time. Considering all this, it can be stated that change management for social aspects is significantly more challenging as it took 30 months to fully implement what was created in 3–4 months. Most of the time was dedicated to people transformation and an effective action plan was needed after scoping.

Industry came to this level of implementation step by step via virtual prototyping. Understanding the link between design, construction, procurement methodology and handover has been crucial, and has required a very long learning curve for the industry. The client should be the managing authority that centralizes the digital twin as the major source of information for all parties.

Further detailing the challenges, it can be said that all parties are the same in terms of managing load since there are different technical problems but a similar mindset is needed for collaboration. Once people experience the benefits of the utilization of digital approaches, it gets easier to manage, but overall no party in the INA project was ready to handle the whole challenge in a fully digitized "one team" mindset because no one was envisioning this output. Consequently, there was a long learning curve for every project party.

7.4 Lean synergies created in the INA project

"How fast did we design and engineer, and how efficiently did we transfer this to manufacturing on site?" is the triggering question. Lean synergies should be the key in this process, and it should not just be for a specific party or interface, but should disseminate through design to installation on site with a design engineering for manufacturing and assembly approach. Consequently, concurrent design and engineering is a very important and topic, and still needs to be tackled in-depth.

The underlying strategy behind achieving lean synergies and concurrency between phases is continuous monitoring of the digital ecosystem, and having full technical knowledge. The party that manages and drives the digital delivery should have a good understanding of technical details. It is a critical responsibility because this is done on behalf of the client. Managing this risk to achieve success leads to many technical challenges, and requires skill to address these challenges in a way that may not please everybody. On the other hand, digital integrated delivery gains the confidence of everybody in such a seamless way, and this helped stop the claims and accelerated the delivery in the INA project.

Specific mechanisms on site are also critical. Mobile BIM is one the backbones that facilitated on site manufacturing and coordination in the INA project. Mobile BIM was the initial strategy from the beginning, but once the design and engineering digital ecosystem needed to be translated into construction site with the mobile utilization of these tools, it became a toolkit of service that pushed the site team to deliver their functionalities on site. In that context, the younger generation are quick learners that take the strategy forward. Overall, a young team can have a huge impact on shortening the learning curve for more experienced staff.

7.5 Lessons learned at the executive level

Similarly, at the executive level of this whole ABIM platform implementation, one needs to deliver in the sense that a mind shift happens. The fragmentation is significant, and experiencing all the different silos and pushing the integration at the executive level is challenging. It is important to initiate merging silos as much as possible. Digital makes sense when the product is out of the door so that it is needed to justify the digital since it is not scientific enough to understand without seeing the actual delivery. This is the reason why AEC is lagging behind in terms of digitization. The easiest way to bring all these parties together is to make them find value in the digital ecosystem, which does not have to be something tangible.

Finally, it is not only about the matter of "digital or digitalization", it is about the mindset. In the future we will be seeing a smaller number of resources with more intelligent operations. Design-engineering-construction ecosystems will eventually be transferred to another ecosystem which is operations. To satisfy the seamless data handover, the available digital environment should be efficient enough to reflect the operational environment requirements. If one does not know what he/she is operating, then it is hard to justify what you are achieving in terms of your KPIs.

The AEC industry has been trying to achieve a transformation in design, engineering and construction for 10–15 years, and the same thing will also be happening for digitization in operations. Artificial intelligence, big data analytics, and more automated workflows will be key drivers to facilitate the digital operations ecosystem in the near future.

Index

9781032570518